Transmission and Processing for Data Center Networking

IOP Series in Advances in Optics, Photonics and Optoelectronics

SERIES EDITOR

Professor Rajpal S Sirohi Consultant Scientist

About the Editor

Rajpal S Sirohi is currently working as a faculty member in the Department of Physics, Alabama A&M University, Huntsville, AL. Prior to this, he was a consultant scientist at the Indian Institute of Science Bangalore, and before that he was Chair Professor in the Department of Physics, Tezpur University, Assam. During 2000–11, he was an academic administrator, being vice chancellor for a couple of universities and the director of the Indian Institute of Technology, Delhi. He is the recipient of many international and national awards and the author of more than 400 papers. Dr Sirohi is involved with research concerning optical metrology, optical instrumentation, holography and speckle phenomena.

About the series

Optics, photonics and optoelectronics are enabling technologies in many branches of science, engineering, medicine and agriculture. These technologies have reshaped our outlook and our way of interacting with each other, and have brought people closer together. They help us to understand many phenomena better and provide deeper insight into the functioning of nature. Further, these technologies themselves are evolving at a rapid rate. Their applications encompass very large spatial scales from nanometers to astronomical and a very large temporal range from picoseconds to billions of years. This series on the advances in optics, photonics and optoelectronics aims to cover topics that are of interest to both academia and industry. Some of the topics that the books in the series will cover include bio-photonics and medical imaging, devices, electromagnetics, fiber optics, information storage, instrumentation, light sources, CCD and CMOS imagers, metamaterials, optical metrology, optical networks, photovoltaics, freeform optics and its evaluation, singular optics, cryptography and sensors.

About IOP ebooks

The authors are encouraged to take advantage of the features made possible by electronic publication to enhance the reader experience through the use of colour, animation and video, and incorporating supplementary files in their work.

Do you have an idea of a book you'd like to explore?

For further information and details of submitting book proposals see iopscience.org/books or contact Ashley Gasque on Ashley.gasque@iop.org.

Transmission and Processing for Data Center Networking

Le Nguyen Binh
Huawei Technologies, European Research Institute, Muenchen, Bayern, Germany

IOP Publishing, Bristol, UK

ISBN 978-0-7503-2292-8 (ebook)
ISBN 978-0-7503-2290-4 (print)
ISBN 978-0-7503-2293-5 (myPrint)
ISBN 978-0-7503-2291-1 (mobi)

DOI 10.1088/978-0-7503-2292-8

Version: 20200202

IOP ebooks

British Library Cataloguing-in-Publication Data: A catalogue record for this book is available from the British Library.

Published by IOP Publishing, wholly owned by The Institute of Physics, London

IOP Publishing, Temple Circus, Temple Way, Bristol, BS1 6HG, UK

US Office: IOP Publishing, Inc., 190 North Independence Mall West, Suite 601, Philadelphia, PA 19106, USA

Contents

Preface

From early in the 2010s, long distance terrestrial and intercontinental telecommunications systems and networks have been reaching toward longer transport distances and increased information capacity through economical implementation of communications systems. These ultra-high capacity long distance transmission systems have been installed throughout the global networks.

Concurrently, the explosive expansion of the Internet drove the tremendous progress in the deployment of data centers (DCs) throughout the worldwide information networks. This has led to long distance and short distance interconnections between these DCs, and hence DC networking without going through traditional telecoms networks. This has put tremendous pressure on global telecoms carriers, as they do not own the content of their traffic, such as the traffic of YouTube, Facebook, Google, etc. However, the infrastructure of these telecoms carriers has allowed them to deploy ultra-high capacity mobile networks, thus the emergence of networks such as 5G (fifth-generation) mobile communications, and thus distributed DC networking. This direction of progress has been strategically developed to challenge the other DC networking communities.

Therefore the transmission of ultra-high speed channels and ultra-high capacity links over short distances has become highly critical for DC networking for both telecoms carriers and data communications for DC network owners. Further, the spectral regions of single-mode silica optical fibers have been extensively used in the O-band (1300 nm window) and C+ (110 channels of 50 GHz spacing), pushing into the L-band, even to the 850 nm region with the low cost vertical-cavity surface-emitting laser (VCSEL). This spectral extension is required to ensure sufficient bands for high capacity transport. The global transport capacity should reach zettabits per second. Thus several possible solutions have been offered by the research and development communities for such capacity demands.

The possibility for such ultra-high speed transmission comes from the digital processing of coherently detected signals, which was not possible in the first generation coherent systems developed in the 1980s. However, when the baud rate of the signals reaches more than 100 GBd, digital signal processing (DSP) faces severe problems with the electronic speed of integrated microelectronics, even in the gate range of less than 7 nm. Optical/photonic signal processing (PSP) can assist in overcoming these difficulties. The main chapters of this book cover the technological principles of these two principal technologies, i.e. DSP-based processing optical systems and PSP systems.

This e-book will thus address the following two principal techniques for the modern communications transmission of ultra-wideband channels:

- DSP-based techniques in optical transmission systems with both non-coherent (direct detection) and coherent principles and associated modulation formats.
- Principles of processing in the photonic domain with special emphasis on the association of neural computing networking and nonlinear optical logic circuits.

This book aims to provide an understanding of such technological developments for practical engineers, R&D engineers, academic scholars, etc.

Acknowledgements

I am grateful to Huawei Technologies for the availability of laboratories and facilities to carry out my research during my time with the company. The chapters of this book were initiated while I was a full-time staff member as a Technical Director of the company European Research Centre in Munich.

I am grateful to Dr Thomas Lee of SHF AG of Berlin, Germany, and formerly of Nortel Networks, for advice and the loan of various optical transmitters and receivers and the purchase of a 64.Gb s^{-1} BERT for several terabit s^{-1} transmission experiments employing various incoherent and coherent M-ary-QAM. I am also indebted to discussions with my colleagues at Keysight Inc. Deutschland and the US Colorado R&D Center, in particular Piotr Laskowski, Thomas Kirchner and Mike Beyers.

I would also like to sincerely thank my former colleagues at Huawei Technologies, in particular Dr Xu Xiaogeng, Bruce Liu, Chen Ming, Dr Mao Bangning, Dr Xie Changsong, Professor Dr Nebojsa Stojanovic, Dr Jinlong Wei, Dr Song Xiaolu, Dr Sun Xu, Dr Suo Jing, Dr Li Rui, Dr Zhao Zhuang and Dr Spiros Mikroulis for fruitful exchanges of ideas and technological information.

Last, but not least, I thank my wife Phuong Nguyen and my son Lam Le Nguyen who have supported me over the years with happiness and for putting up with the many hours spent in the preparation of this book. My parents, the Le-Nguyen family, have provided me over the years with an education in 'learning for life' and I am indebted to them for their sacrifices in educating their children and their life-long encouragement. This book is dedicated to my parents.

Le Nguyen Binh, PhD, DrEng, CEng
Principal Expert Consultant for Huawei Technologies,
European Research Institute, München, Bayern, Germany.
September 2019

Author biography

Le Nguyen Binh

Dr DrEng Professor Le Nguyen Binh received his BE in electronic engineering (Summa cum Laude) and PhD, DrEng in electronic engineering and integrated photonics/optical communications engineering, respectively, from the University of Western Australia, Nedlands, WA. He is a Chartered Engineer. He has published more than 300 journal and conference papers, owns several patents and has published 14 books in the fields of optical communications transmission systems and network engineering, photonic processing, integrated optics and electromagnetics engineering.

Abbreviations

Abbreviation	Expansion
ADC	analog to digital converter
ASK	amplitude shift keying
C-band	spectral region 1550 nm (40 nm window 1525–1565 nm C+ 1580 nm)
CD	chromatic dispersion
CDC	cloud data center
Co-OFDM	coherent OFDM
CONN	convolutional optical neural network
CPFSK	continuous phase frequency shift keying
DAC	digital-to-analog converter
DC	data center
DDC	distributed data centers
DD-OFDM	direct detection OFDM
DCI	DC interconnection
DCN	DC networking
DCF	dispersion compensating fiber
DFB	distributed feedback
DFE	decision feedback equalizer
DMT	discrete multi-tone
DPSK	differential phase shift keying
DSP	digital signal processing
DWDM	dense WDM
EDFA	erbium-doped fiber amplifier
ENOB	effective number of bits
FEC	forward error coding
FFE	feed forward equalizer/equalization
FFT	fast Fourier transform
FT	Fourier transform
Gbps	gigabits s^{-1}
IFT	inverse Fourier transform
IFFT	inverse FFT
IT	information technology
ITU	International Telecommunications Union
LO	local oscillator
M-ary-QAM	multi-level QAM
Mbps	megabits s^{-1}
MCU	micro-control unit
MLSE	maximum likelihood sequence estimation
MMF	multi-mode fiber
MSK	minimum shift keying
MZDI	Mach–Zehnder delay interferometer
MZ(I)M	Mach–Zehnder (interferometric) modulator
O-band	spectral region 1300 nm (35 nm window)
OA	optical amplifier
OCNN	optical computing neural network
O/E	optical to electrical conversion
OFDM	orthogonal frequency division multiplexing

OFDR	optical frequency domain reflectometry
OFDR	optical frequency discrimination receiver
OIF	Optical Internet Working Forum
OLG	optical logic gate
ONN	optical neural network
OOK	on–off keying
OFT	optical FT
ORNN	optical reservoir neural network
PAM	pulse amplitude modulation
PMD	polarization mode dispersion
PSP	photonic signal processing
QAM	quadrature amplitude modulation
QPSK	quadrature phase shift keying
RCNN	reservoir neural computing network
SFG	small form-factor pluggable
SMF	single-mode fiber
SPM	self-phase modulation
SSMF	standard SMF
Tbps	terabits s^{-1}
TIA	trans-impedance amplifier
TO	thermal optic
TV	television
WDM	wavelength division multiplexing
XPM	cross phase modulation
Zbps	zettabits s^{-1}

IOP Publishing

Transmission and Processing for Data Center Networking

Le Nguyen Binh

Chapter 1

Historical overview and digital transmission technologies in cloud networking

Over the last half century, optical guided media, the circular optical waveguide and the optical fiber have undoubtedly allowed tremendous progress in transporting huge amounts of information. This introductory chapter briefly outlines the historical development, emergence and merging of the fundamental digital communication techniques and optical communications to fully exploit and respond to the challenges of the availability of the ultra-high frequency and ultra-wide bands in the optical spectra of optical fiber communications technology. Currently, with the emergence of cloud networking (where the contents are owned by entertainment and social media delivery corporations such as YouTube, Netflix, Facebook, etc) coupled with communications platforms, networking has evolved into ultra-high capacity, both in the core and distributed data center clouds (DCCs), and in terms of both intra- and inter-networking. This has forced the evolution of the traditional telecoms carrier networks to adapt to DCC networking, taking advantage of their existing infrastructure. Wireless access networks have evolved into 5G and this has forced the distribution of the cloud into distributed data centers (DDCs), local nodes of 5G networking. The extensive infrastructure of telecoms networks gives them tremendous leverage over the data center (DC) network owners. However, the delivery and processing of signals and channels over these networks are critical for error free transmission to be achieved.

The processing of these ultra-wideband channels is critical and is reaching the limit in the electronic domain, but has advantages in the photonic domain. This chapter thus provides introductory remarks and a basic understanding for the remaining chapters of the book. A historical overview and the emergence of the challenges of processing in optical communications are provided to emphasize the importance of the field.

This chapter introduces a historical overview of optical transport through optical guided media, from the innovative ideas of 1966 to the beginning of electronic digital

doi:10.1088/978-0-7503-2292-8ch1

signal processing (DSP) based optical transmission and then to the DSP based systems. We also cover the growth of traffic via DC networking and the demand for extreme data rate transmission and interconnection, reaching an approximation of terabits s^{-1} (Tbps) per wavelength carrier. The basic modulation technique for such signal generation, in both amplitude and phase modulation, as well as the frequency division modulation with orthogonality are given. Then, electronic DSP based optical systems and photonic signal processing based systems are introduced to provide basic information for the remaining chapters of this book.

1.1 Non-DSP based optical transmission technologies

Starting from the dielectric waveguide proposed by Kao and Hockham [1, 2] in 1966, the first research phase attracted intensive interest around the early 1970s in the demonstration of fiber optics, and optical communications have progressed tremendously over the last five decades. The first-generation lightwave systems were commercially deployed in 1983 and operated in the first wavelength window of 800 nm over multi-mode optical fibers (MMF) at transmission bit rates of up to 45 Mb s^{-1} [1]. After the introduction of ITU-G652 standard single-mode fibers (SSMF) in the late 1970s [3], the second generation of lightwave transmission systems became available in the early 1980s [4, 5]. The operating wavelengths were shifted to the second window of 1300 nm, which offers much lower attenuation for silica-based optical fibers compared to the previous 850 nm spectral window region; in particular, the chromatic dispersion (CD) factor is almost zero. These second generation systems could operate at bit rates of up to 1.7 Gb s^{-1} and have a repeaterless transmission distance of about 50 km [6]. Further research and engineering efforts were also aimed at the improvement of receiver sensitivity through coherent detection techniques, and the repeaterless distance reached 60 km in installed systems with a bit rate of 2.5 Gb s^{-1}. Optical fiber communications then evolved to third-generation transmission systems, which utilized the lowest attenuation 1550 nm wavelength window and operated at an up to 2.5 Gb s^{-1} bit rate [7]. These systems became commercially available in 1990 with a repeater spacing of around 60 to 70 km [8]. At this stage, the generation of optical signals was mainly based on direct modulation of the semiconductor laser source and either direct or coherent detection. Since the invention of erbium-doped fiber amplifiers (EDFAs) in the early 1990s [9–11], lightwave systems have rapidly evolved to wavelength division multiplexing (WDM) and, shortly after, dense WDM (DWDM) optically amplified transmission systems, which are capable of transmitting multiple 10 Gb s^{-1} channels. The 10 Gbps systems are dominated by Nortel Networks Inc., and made the decreasing market share of 2.5 Gbps the main line of Lucent Technologies. This is due to the fact that loss is no longer a major issue for external optical modulators that normally suffer an insertion loss of at least 3 dB. These modulators allow the preservation of the narrow linewidth of distributed feedback (DFB) lasers. These high speed and high capacity systems extensively exploited external modulation in their optical transmitters. The current optical transmission systems are considered as the fifth generation, having transmission capacities of a few tens of Tb s^{-1}.

Coherent detection, homodyne or heterodyne and intradyne, was the focus of extensive research and development (R&D) during the 1980s and early 1990s [12–17], and was the main detection technique in the first three generations of lightwave transmission systems. At that time, the main motivation for the development of coherent optical systems was to improve the receiver sensitivity, commonly by 3–6 dB. The repeaterless transmission distance was thus able to be extended to more than 60 km of SSMF (with an 0.2 dB km^{-1} attenuation factor). However, coherent optical systems suffer severe performance degradation due to fiber dispersion impairments. In addition, the phase coherence for lightwave carriers of the laser source and the local laser oscillator was very difficult to maintain. In contrast, the incoherent detection technique minimizes the linewidth obstacles of the laser source as well as the local laser oscillator and thus relaxes the requirement for phase coherence. Moreover, incoherent detection mitigates the problem of polarization control in the mixing of transmitted lightwaves and the local laser oscillator in the multi-THz optical frequency range. The invention of EDFAs, which are capable of producing optical gains of 20 dB and above, has also greatly contributed to the progress of incoherent digital photonic transmission systems to date.

In recent years there has been a huge increase in the demand for broadband communications, mainly driven by the rapid growth of multi-media services, peer-to-peer networks and IP streaming services, in particular IP TV. It is most likely that such tremendous growth will continue in the coming years. This is the main driving force for local and global telecommunications service carriers to develop high performance and high capacity next-generation optical networks. The overall capacity of WDM or DWDM optical systems can be boosted either by increasing the base transmission bit rate of each optical channel, multiplexing more channels in a DWDM system or, preferably, by combining both of these schemes. However, while implementing these schemes, optical transmission systems encounter a number of challenging issues as outlined in the following (figure 1.1).

The current extensively deployed 100 Gbps transmission systems employ intensity modulation (IM), also known as on–off keying (OOK) and utilize non-return-to-zero (NRZ) pulse shapes via either direct detection or coherent receiving sub-systems using four wavelength carriers or over four lanes (figure 1.1). The term OOK can also be used interchangeably with amplitude shift keying (ASK)[1]. Moving toward high bit rate transmission such as 100 Gbps, the performance of OOK photonic transmission systems is severely degraded due to fiber impairments, including fiber dispersion and fiber nonlinearities. The fiber dispersion is classified into chromatic dispersion (CD) and polarization mode dispersion (PMD), causing inter-symbol interference (ISI) problems. Severe deterioration of the system's performance due to fiber nonlinearity results from high power spectral components at the carrier and signal frequencies of OOK-modulated optical signals. It is also of concern that existing transmission networks comprise millions of kilometers of

[1] The OOK format simply implies the on–off states of the lightwaves where only the optical intensity is considered. In contrast, the ASK format is a digital modulation technique representing the signals in the constellation diagram by both the amplitude and phase components.

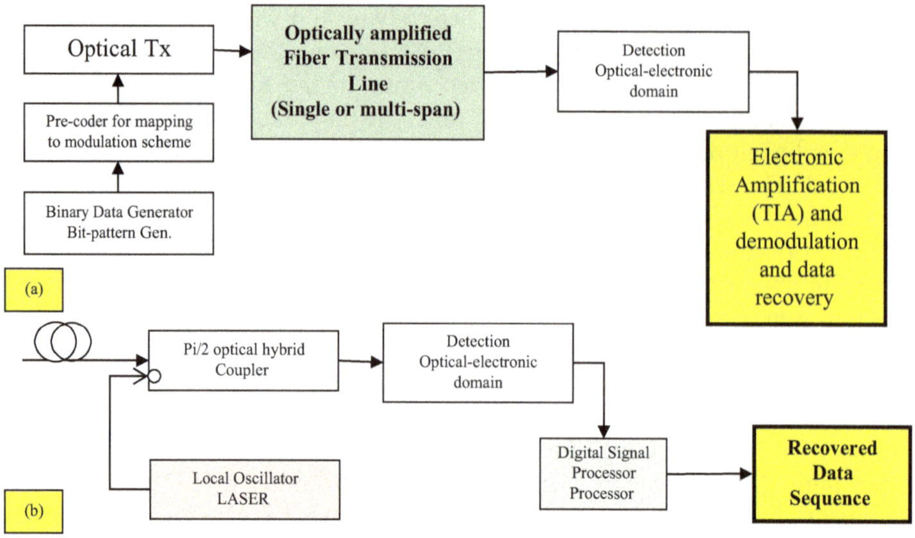

Figure 1.1. Schematic diagram of the modulation. (a) Electronic detection and demodulation of an advanced modulation format optical communications system. (b) Modulation and direct detection if no local oscillator laser is employed. Optical transmission link in single- or multi-cascade spans.

SSMF, which has been installed for approximately two decades. These fibers do not have as advanced properties as the state-of-the-art fibers used in recent laboratory 'hero' experiments and they have degraded after many years of use.

The total transmission capacity can be enhanced by increasing the number of multiplexed DWDM optical channels. This can be carried out by reducing the frequency spacing between these optical channels, for example from 100 GHz to 50 GHz, or even 25 GHz and the even narrower 12.5 GHz [18, 19]. The reduction of the channel spacing also results in narrower bandwidths for optical multiplexers (mux) and demultiplexers (demux). Passing through these narrowband optical filters, the signal waveforms are distorted and optical channels suffer the problem of inter-channel cross-talk. The narrowband filtering problems are becoming more severe at high data bit rates, e.g. 56 GBd and even 112 GBd, thus degrading the system performance significantly.

Together with the demand for boosting the total system capacity, another challenge for service carriers is to find cost-effective solutions for the upgrading process. These cost-effective solutions should require minimum renovation to the existing photonic and electronic sub-systems, i.e. the upgrading should only take place at the transmitter and receiver ends of an optical transmission link. Another possible cost-effective solution is to significantly extend the uncompensated reach of optical transmission links, i.e. without using dispersion compensation fibers (DCF), thus reducing by a considerable number the required inline optical amplifiers, the EDFAs. This network configuration has recently received much interest from both the photonics research community as well as service carriers.

1.2 Signal processor DSP based optical transmission

The bit rate carried by each optical carrier can be increased by improvement of the narrowness of the linewidth of the lasers as the light source and a local oscillator with reasonably high power, ~40 mW in the 1550 nm window. Furthermore, the advancement of digital signal processors (DSPs) in microelectronic technology, together with the ultra-high sampling rate via analog-to-digital converters (ADCs) and digital-to-analog converters (DACs) of the received optical signals in continuous form sampled at very high bit rates by digital processors, permit the compensation of fiber impairments and phase recovery of system clocking as well as the determination of the frequency offset of the local oscillator (LO) laser and that of the optical carrier. Hence, these approaches can overcome the severe problem of homodyne reception systems faced by the first generation of coherent transmission systems in the early 1980s.

Over the last few decades, extensive R&D has demonstrated coherent reception incorporating a DSP that can push the bit rates per wavelength channel to 100 Gb s^{-1} and then 800 G and beyond, employing 25 G/100 GBd polarization multiplexing and QPSK (2 bits/symbol) and/or M-ary QAM to aggregate to 200 Gbps or 400 Gbps/800 Gbps. Furthermore, the channels can be pulse shaped by using DACs to pack the channels into super-channels to generate Tbps per channel with sub-carriers. The ultra-high sampling rate ADCs and DACs have reached 64 GSa s^{-1} to 200 GSa s^{-1}, which allow the DSP to recover the clock, hence the sampling rate and time, combating the linear and nonlinear impairments due to CD, PMD, self-phase modulation (SPM), cross-phase modulation (XPM) and other effects. The transmission for 100 Gbps would reach 3500 km in a field trial and 1750 km for 200 Gbps over an optically amplified and a non-DCF span transmission distance. However, the power consumption is extremely high, more than 15 W for a 200 GSa s^{-1} sampling rate. It is noted that the power consumption of a DSP is increased as a cubic exponential with respect to the increasing factor of the sampling rate.

In addition to coherent reception, the use of DSP is also applied to IM/DD transmission of signals, which can take modulation formats such as PAM-*n* [20] or discrete multi-tone (DMT) [21–23] or orthogonal frequency division multiplexing (OFDM). DMT is indeed OFDM for intensity modulation and direct detection (IMDD) transmission.

1.3 Increasing transmission capacity over short distances

Capacity constantly needs to be increased, and in early 2010 multi-core fibers (MCFs) were demonstrated in which a common cladding and several single-mode cores are embedded. Further optical amplifiers were developed via cladding pumping for structuring optically amplified MCF spans for medium distance transmission links.

However, when the baud rate reaches 112 GBd and beyond, the power consumption and electronic speed of the DSP create severe difficulties. Thus, by processing the signals in the photonic domain, photonic signal processing (PSP) offers significant advantages compared to DSP in the electronic domain. This is one of the main topics of this book. PSP is a processing technique in the analog domain without memory or at most the memory length is of one.

The transmission distance and bit rate are evolving based on the networking demands. Cloud and DC networking are evolving at a very fast pace. Furthermore, 5G networks are developing and are expected to be extensively deployed from 2020 onwards, demanding extreme capacities of accessing devices and local clouds. In addition, the intra-DC interconnections require even higher rates and all-optical interconnection and switching. The transmission technologies are thus evolving to meet these demands, i.e. ultra-high rates with shorter distances with minimum latency. The next section addresses this technological evolution.

1.4 Digital optical transmission in evolving networking

1.4.1 Traffic growth

It is astounding that the growth of telecommunications traffic never stops. We have observed this since 1966, the year the idea of guiding light over a dielectric waveguide was invented, with a bit rate of a few Mbits s^{-1} over only a few kilometers of multi-mode fibers (MMF). To date the distance has become 2 km to 80 km for intra-DC transport to metro access, and the bit rate can reach 800 Gbps or even beyond. The demands on data transport over the Internet have exploded due to the resources required for big or giant data pipelines, etc. Data storage and delivery on demand through the DCs or clouds is required for a mixed data communication of traditional telecoms and cloud networking for 5G and the Internet of Things (IoT). The total capacity of information to be transported can reach zettabits per second (figure 1.2).

The growth in the communication of humans and machines to machines and cloud computing drives the growth of information pipelines to become bigger and bigger, as shown in figure 1.3. The expected evolution of the transport network is shown in the schematic of figure 1.4, where the demands in large bandwidths can reach multi-Tbps per optical carrier. So how can the telecoms carriers and data networking corporations evolve to meet the common networking requirements in the near future for the benefit of global communities?

A market overview of the transport technology and total bit rate per transmission link is illustrated in figure 1.5, for the 100 G and the emerging 400 G market. Furthermore, 800 G and Tbps are also emerging to supply the demands of traffic transport, as observable in figure 1.4. This figure gives the time-line of standards developed for 100 G and 400 G by international bodies. The 400 G standard was already issued in 2017. This has put a high importance on this rate and those beyond.

These data rates are increased in order to supply capacity for the evolving data center networking with accommodating for growth, in particular for 5G mobile services. The evolutionary structures of DC and emergence of modified DC distribution structure can be seen in figure 1.6.

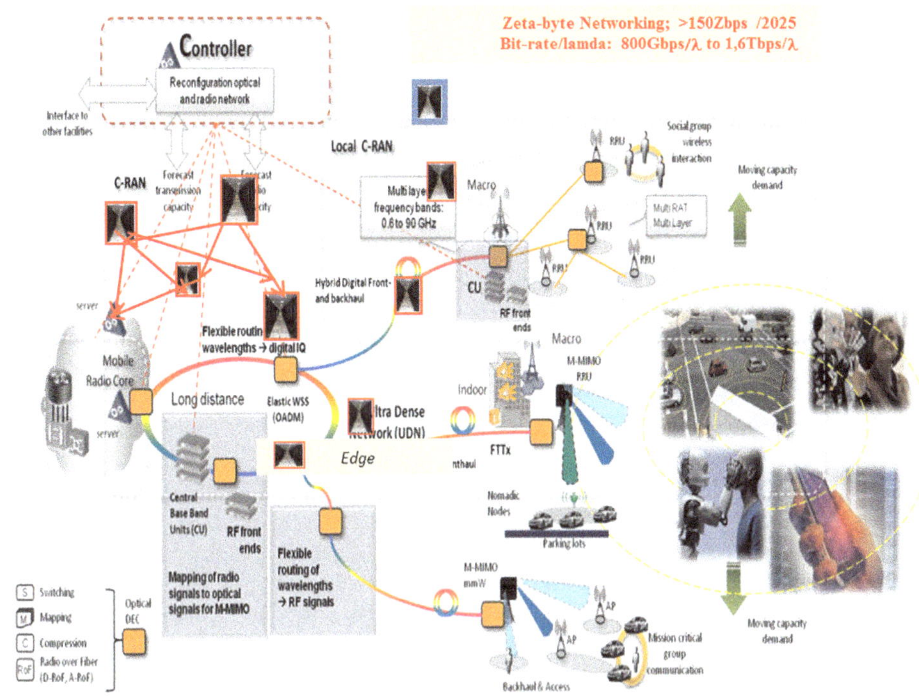

Figure 1.2. DC distribution in mixed traditional telecoms and data center networking in expected 5G network scenarios.

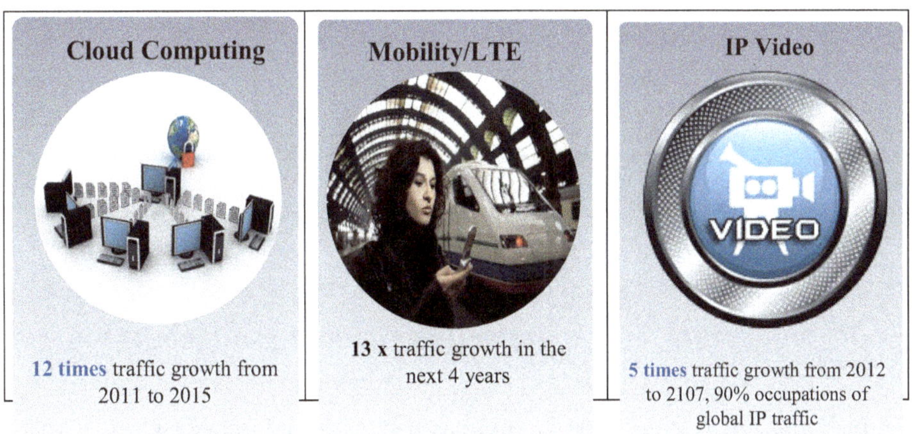

Figure 1.3. The explosive growth of traffic (in growth rate) in cloud computing, mobile communications and internet protocol (IP) video drives the need for larger bandwidth pipelines and machine-to-machine 5G.

1.4.2 Transmission technologies for intra- and inter-DC cloud networking

Data centers have evolved significantly in recent years, adopting technologies such as virtualization to optimize resource utilization and increase IT flexibility. As an enterprise IT needs to continue to evolve toward on-demand services, many

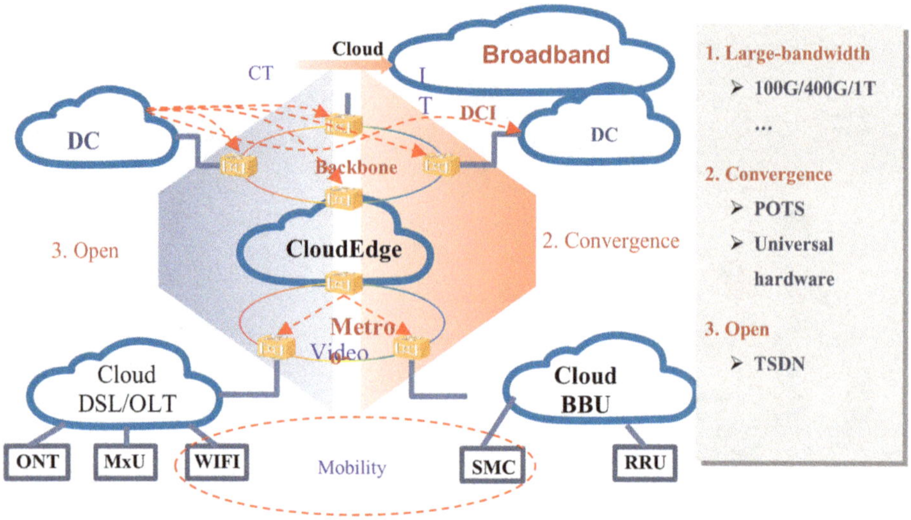

Figure 1.4. A near-future transport network under software control architecture from broadband to distribution of clouds and DC networking. RRU = radio remote unit; B = broadband; DCI = DC interconnect; WI-FI = wireless fiber; DSL = digital service line; OLT = optical line transport; ONT = optical transport unit; MxU = multiple cross unit; SMC = service mobile control; TDSN = time division data network.

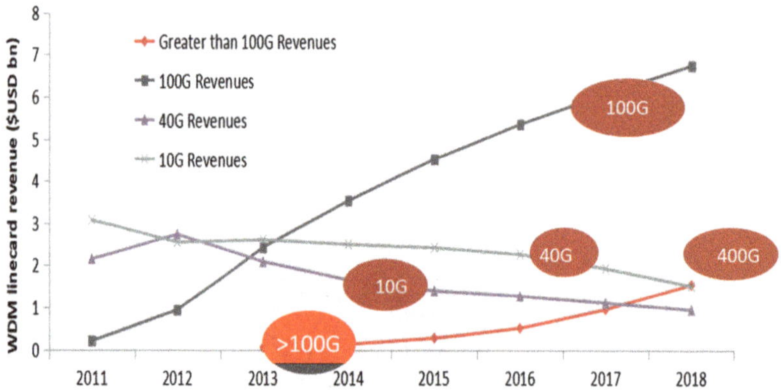

Figure 1.5. Market overview: 100 G is surging and 400 G is emerging with expected exponential growth in the next decade.

organizations are moving toward cloud-based services and infrastructure. Focus has also been placed on initiatives to reduce the enormous energy consumption of DCs by incorporating more efficient technologies and practices in data center management. DCs built to these standards have been termed 'green DCs'. The interconnections between servers and interconnections between DCs are critical for communications and serving demands. Thus the technological understanding of these transmission links is the main subject of this book.

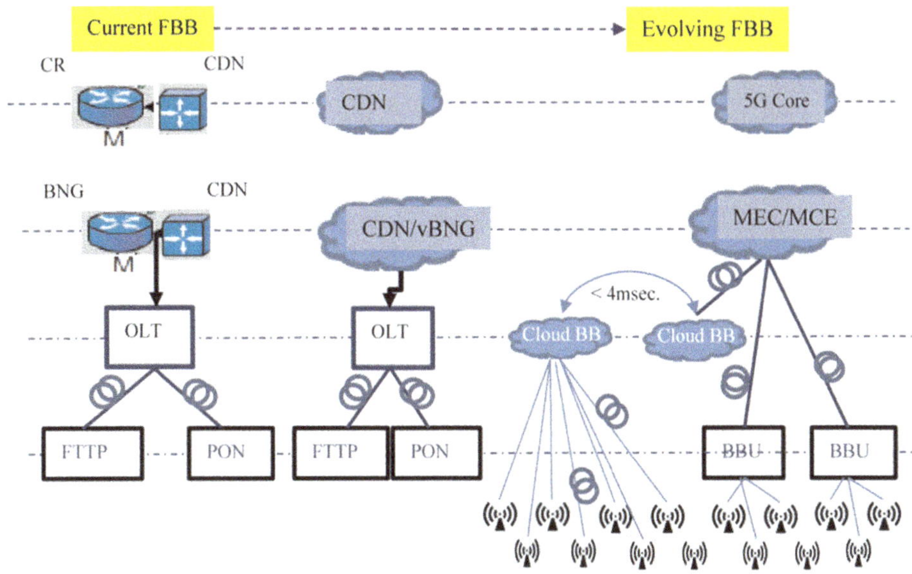

Figure 1.6. Evolution of current transport to cloudified networks for 5G. CDN = content delivery networks; FBB = fixed broadband; CDN = current delivery network; vBNG= variable business network group; OLT = optical line terminal; MCE = mobile cloud engine; BBU = baseband unit; FTTP = fiber to the premise; PON = passive optical networks. All optical paths are fiber based cabling.

Figure 1.7. CERN data center. Image source: Hugo van Meijeren.

1.4.3 DCs and cloud computing—a historical overview

A DC is a facility used to house computer systems and associated components, such as telecommunications, storage and server systems (figure 1.7). It generally includes redundant or backup power supplies, redundant data communications connections, environmental controls (e.g. air conditioning, water cooling and fire suppression) and various security devices. A large DC is an industrial-scale operation using as much electricity as a town. Some DCs can consume aggregate power which is equivalent to that generated from a medium scale nuclear power station, or about a few hundreds of megawatts per hour.

DCs have their roots in the huge computer rooms of the 1940s, one of the earliest examples of a DC is shown in figure 1.8. Early computer systems were complex to operate and maintain, and required a special environment in which to operate, the computer center. Many cables were necessary to connect all the components, and methods to accommodate and organize these were devised, such as standard racks to mount equipment, raised floors and cable trays (installed overhead or under the elevated floor). A single mainframe required a great deal of power and had to be cooled to avoid overheating.

During the boom years of the microcomputer industry, and in particular during the 1980s with the operating speed reaching several GHz and sampling rates of a few hundreds of GSa s^{-1}, users started to deploy computing systems everywhere, in

(a)

(b)

Figure 1.8. (a) Very classical manual switching and interconnection of telephone systems (circa 1950s–1960s); courtesy of US Department of Energy. (b) Initial digital automatic switching and backplane cabling system; source: Midom, Flickr.

many cases with little or no care for the operating requirements. However, as IT operations started to grow in complexity, organizations grew aware of the need to control IT resources. The advent of Unix from the early 1970s led to the subsequent proliferation of freely available Linux-compatible PC operating systems during the 1990s. These were called ‘servers’, as sharing operating systems such as Unix relies heavily on the client–server model to facilitate sharing unique resources between multiple users. The availability of low-cost networking equipment, coupled with new standards for network structured cabling, made it possible to use a hierarchical design that put the servers in a specific room inside the company. The use of the term ‘data center’ (DC), as applied to specially designed computer rooms, gained popular recognition.

The boom of DCs came during the dot-com bubble of 1997–2001. At this time the transmission of 10 Gbps and 40 Gbps using a $LiNbO_3$ modulator in transmitter and self-homodyne detection without any DSP based processing in the receiving system was achieved. Companies needed fast Internet connectivity and non-stop operation to deploy systems and to establish a presence on the Internet. Installing such equipment was not viable for many smaller companies. Many companies started building very large facilities, called Internet DCs (IDCs), which provided commercial clients with a range of solutions for systems deployment and operation. New technologies and practices were designed to handle the scale and the operational requirements of such large-scale operations. These practices eventually migrated toward private DCs, and were adopted largely because of their practical results. DCs for cloud computing are called cloud DCs (CDCs). Nowadays, the division of these terms has almost disappeared and they are being integrated into the general term ‘data center’ (DC).

With the increased employment of cloud computing, business and government organizations began to scrutinize DCs to a higher degree in areas such as security, availability, environmental impact and adherence to standards. Standards documents from accredited professional groups, such as the Telecommunications Industry Association (TIA), specify the requirements for DC design. Well-known operational metrics for DC availability can serve to evaluate the commercial impact of a disruption. Development continues in operational practice, and also in environmentally friendly DC design. DCs typically cost a lot to build and to maintain and have high power consumption.

1.4.3.1 Requirements for modern DCs

IT operations are a crucial aspect of most organizational operations around the world (figure 1.9). One of the main concerns is business continuity; companies rely on their information systems to run their operations. If a system becomes unavailable, company operations may be impaired or stopped completely. It is necessary to provide a reliable infrastructure for IT operations in order to minimize any chance of disruption. Information security is also a concern. For this reason a DC has to offer a secure environment which minimizes the chances of hacking. A DC must therefore maintain high standards to assure the integrity and functionality of its hosted computer environment. This is accomplished through redundancy of

Figure 1.9. Backplane interconnection of a moderate server system, even so massive cabling. Source: Dave Herholz, Flickr.

mechanical cooling and power systems (including emergency backup power generators) serving the DC along with fiber-optic cables (figure 1.9).

The TIA's Telecommunications Infrastructure Standard for DCs specifies the minimum requirements for the telecommunications infrastructure of DCs and computer rooms, including single-tenant enterprise DCs and multi-tenant Internet hosting DCs. The topology proposed in this document is intended to be applicable to a DC of any size.

The DC equipment may be used to: (i) operate and manage a carrier's telecommunication network; (ii) provide DC based applications directly to the carrier's customers; (iii) provide hosted applications for a third party to provide services to their customers; and (iv) provide a combination of these and similar DC applications.

Standardization and modularity can yield savings in the design and construction of telecommunications DCs. Modularity is important for scalability and thus easier low-cost growth, even when planning forecasts are less than optimal. For these reasons, telecommunications DCs should be planned in repetitive building blocks of equipment, and associated power and support (conditioning) equipment when practical. The use of dedicated centralized systems requires more accurate forecasts of future needs to prevent expensive over construction, or perhaps worse—under construction that fails to meet future needs.

The 'lights-out' DC is a DC that, ideally, has all but eliminated the need for direct access by personnel, except under extraordinary circumstances. All of the devices are accessed and managed by remote systems, with automation software controlling operations for energy savings, hence reducing operational expenditure costs and providing the ability to locate the site further from population centers.

There is a trend to modernize DCs in order to take advantage of the performance and energy efficiency increases of newer IT equipment and capabilities, such as cloud computing. This process is also known as DC transformation.

Organizations are experiencing rapid IT growth but their DCs are aging. The industry research company International Data Corporation (IDC) puts the average

age of a DC at nine years old. The growth in data (163 zettabytes by 2025) is one factor driving the need for DCs to modernize. The typical projects within a DC transformation initiative include standardization or consolidation, virtualization, automation and security.

1.4.3.2 Virtualization

There is a trend to use IT network function virtualization (NFV) technologies to replace or consolidate multiple DC devices, such as servers. Virtualization helps to lower capital and operational expenses, and reduce energy consumption. Virtualization technologies are also used to create virtual desktops, which can then be hosted in DCs and rented out on a subscription basis.

1.4.3.3 Automation

DC automation involves automating tasks such as provisioning, configuration, patching, release management and compliance. As enterprises suffer from a lack of skilled IT workers, automating tasks makes DC operation more efficient. Software defined networking (SDN) in the optical domain or in the software environment are critical for automatic control centers to reconfigure the DC networking and linking to other DCs.

1.4.3.4 Security

In modern DCs, the security of data in virtual systems is integrated with the existing security of the physical infrastructure. The security of a modern DC must take into account physical security, network security, and data and user security. Physical and software security are both critical. Quantum key distribution (QKD) is now reaching a state of absolute security.

1.4.3.5 Carrier neutrality

Currently, many DCs are run by Internet service providers solely for the purpose of hosting their own and third party servers. However traditionally DCs were either built for the sole use of one large company, or as carrier hotels or network-neutral DCs. These facilities enable the interconnection of carriers and partners, and act as regional fiber hubs serving local businesses in addition to hosting content servers.

1.4.4 100 G, 400 G and beyond—transmission technologies for cloud networking

In brief the 100 G module consists of four lanes (wavelength) and each lane carries 25 Gbps. The inter-link between servers in a DC requires the size and power consumption to be as small as possible. The CFP[2] to CFP2 and then CFP4 are preferable for size, as illustrated in figure 1.10, for a 100 G module from a client access link or in DC inter-sever connections. Some technical details are described in chapter 4. Figure 1.11 gives the timelines of the standardization for 400 G and 100 G. The standard 100 G has already been issued and was developed quickly for

[2] CFP module = C form-factor pluggable module; C = centum (100).

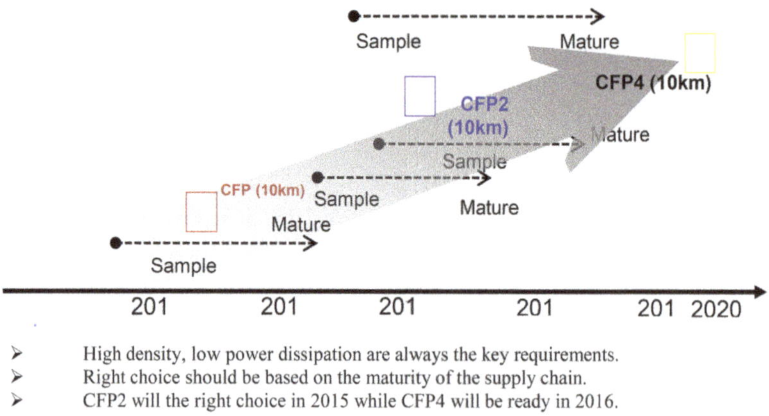

Figure 1.10. 100 G client side module evolution with the physical size from CFP to CFP2 and then CFP4 with the availability of samples for testing and maturity.

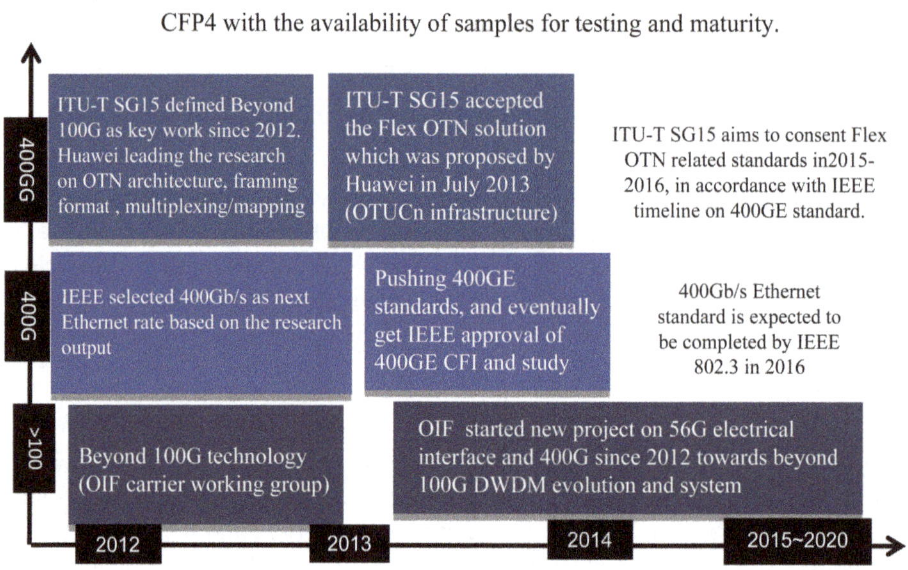

Figure 1.11. Standard timelines for 200 G and 400 G. OIF = optical interchange forum; IEEE = Institute of Electrical and Electronic Engineers.

several deployments in DC networking and transmission[3,4]. This is to ensure that multi-vendor modules can be used, thus lowering the cost of these plug and play approaches. The 400 G rate is also based on the modulated channels on four lanes but with 56 GBd and PAM4 (2 bits/symbol) to give 100 G per lane. Thus four lanes would aggregate to 400 G.

---

[3] IEEE Standard 802.3ba: Media Access Control Parameters, Physical Layers, and Management Parameters for 40 Gb/s and 100 Gb/s Operation. Approved by IEEE and ANSI on 17 June 2010, http://standards.ieee.org/findstds/standard/802.3ba-2010.html.

[4] IEEE P802.3bs 400 GbE Task Force, http://www.ieee802.org/3/bs/public/14_05/index.shtml.

1.5 Photonic signal processing

1.5.1 Motivations and backgrounds

The transmission of data channels has been modernized in the following ways:

- Long distance and high bit rate under a coherent technique. Low-cost coherent transmission modules for short distances with ultra-high bit rate interconnections.
- For short distance DC transmission, DC networking (DCN) and DC interconnection (DCI) inter-channel transmission using coherent and IM/DD with the bit rate reaching 800 Gbps and 250 Gbps, respectively. The basic baud rate can now reach 112 GBd and even higher
- DSP with a sampling rate reaching 200 Gbps is needed. This creates a lot of strain on the electronics [24].

Thus the processing of signals in the photonic domain will avoid this constraint and will provide a much higher rate due to the high speed sampling in the optical domain. For a 1.0 ps sampling time it requires a waveguide delay length of only about 120 μm. Thus for a 10 ps pulse period, sampling ten times 1.0 tera-samples per second presents no difficulty[5].

Optical neural network computing has also been a topic of interest since 1990. However, its use and application in transmission has not been investigated. Neural computing allows the distribution of field and intensity and their weighting to hidden layers of neurons so that several effective algorithms of computing can be processed. This is a very powerful computing technique.

We propose here, for the first time, the great potential of using optical neural computing (ONC) for processing of signals in the optical domain, photonic signal processing (PSP). ONC takes in the optical samples from the optical sampler, and then distributes them to other neurons with variable weighting. Optical neurons are designed using multi-mode interference (MMI) devices, incorporating an optical phase shifter to vary the output field strength of the MMI outputs. Thus MMI acts like optical neurons [25, 26].

Recently published articles on optical logic circuits employing nonlinearity of phases in a semiconductor optical amplifier (SOA) have given us a new look at PSP [27], because such logic would assist in the optical decision making for the recovery of ‘high’ or ‘low’ states, thus in data transmission systems.

The availability of optical logic gates (OLGs) determines the final decision that can be obtained by a summation of all optical branches inserting into the input nodes or making a comparison logically to output an amplified and saturated optical signal. The OLG is structured with an SOA embedded in the branches of an MZI, the nonlinearity in-phase of the SOA offers many nonlinear properties which are important for optical decisions or optical logical functions (see figure 1.12).

[5] Note: for 1.0 ps delay. Under a Si on insulator (SOI) integrated photonic waveguide, the length required is 120 μm as the effective ref index is 2.55. Thus for ten optical samples or slicing of the optical signals the total length of the samples would be of the order of 1.2 mm long.

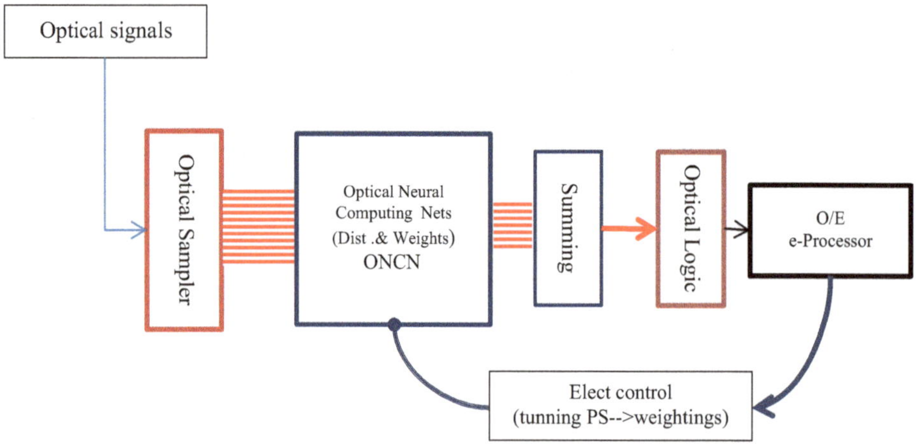

Figure 1.12. Generic model of a time-domain PS processor with electrical feedback to control the weightings of ONCN.

Therefore, our innovative aspects are thus based on an optical sampling structure such that the optical signal can be sliced into a number of optical sampled slices, which are then fed into the ONCN (OPtical Neural Computing Networks) with a distribution whose weighting coefficients can be tuned by optical phase shifters, the optical neurons, which are named for the first time here. The neural computing algorithms are performed in the optical domain, and coefficients or weighting are tunable via thermal tuning of phase shifters on branches of inputs to optical neurons (see figure 1.13). Then a summation of all outputs of the hidden intermediate neuron layers is sent to a decision neuron whose output is fed into the OLG so that a 'high' or 'low' optical state is obtained. It is then straightforward from this point to provide electrical recovery signals.

In formulating the photonic processors, we base our various structures on the optical sub-modules as described above, such as the detection of deferentially coded optical sequences and forward error coding (FEC). Several other processing algorithms can be developed from this fundamental principle.

Our proposal, therefore, gives a detailed account of various novel structures of PSP but demonstrates only one basic PSP structure using the three basic optical sub-modules as briefly described above and in the following. It is believed that such a PSP as proposed here will lead to several patents and lead the global R&D on photonic processors, in particular for telecommunications systems for DCN and DCI.

1.5.2 Generic innovative models of photonic signal processors

This section outlines the generic models for PSPs so as to give the distinctions of our models in dealing with optical channels, as follows:

- The PSP model in the time domain without optical feedback, but electrical control feedback whose electrical signals are derived from the output of the processing system.

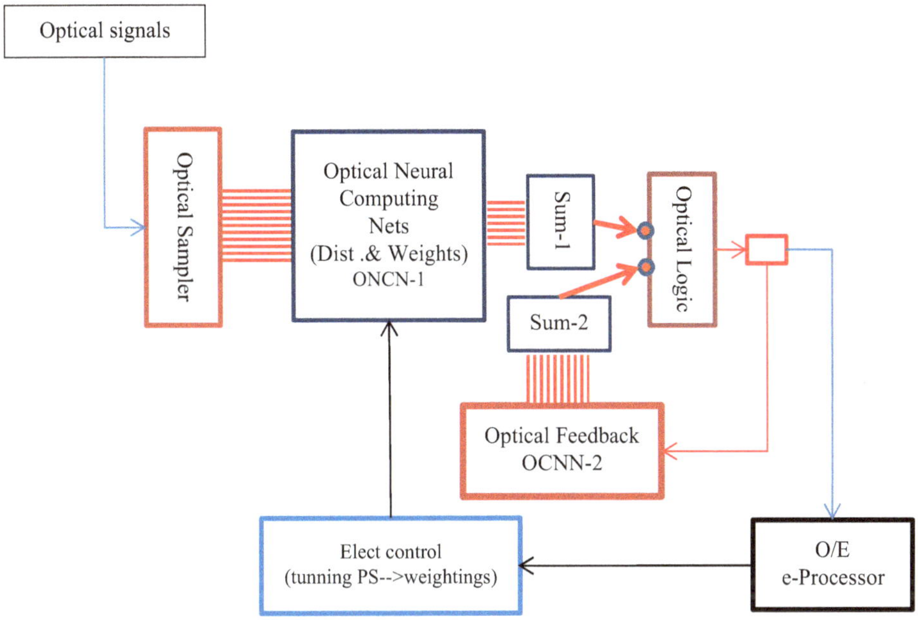

Figure 1.13. Generic model of time-domain PSP with optical feedback to OCNN-2 to compare with OCNN-1 by one bit (or longer) delay in the optical logic circuit plus e-feedback control of the weightings of ONCN-1.

- The PSP model in the time domain with optical feedback.
- The PSP model in the spectral domain.

In principle, we can see the operational points. In the case of temporal domain PSP:

- The optical signals are converted into samples and from serial-to-parallel (S/P) conversions by the use of optical samplers (see figure 1.12).
- The optical samples are then processed in an optical computing neural network (OCNN) by distribution into intermediate layers (the hidden layers) whose paths are weighted by tuning the coupling coefficients of the optical couplers (e.g. multi-mode interferometric coupler).
- The outputs of the OCNN are then summed to give a resulting optical output which is then fed into an optical logic circuit whose output gives a 'high' or 'low' amplified optical signal which is then converted into a recovered electrical signal.
- The electronic processor for this PSP can be much simpler and possibly as simple as a low speed micro-controller unit (MCU) for controlling and tuning of the optical section via thermo-optic (TO) effects.

In the case of an optical feedback time-domain PSP (see figure 1.13), the optical output of the optical logic circuit is tapped and then further processed via a second OCNN whose delay time can be within one symbol period or longer as desired. The

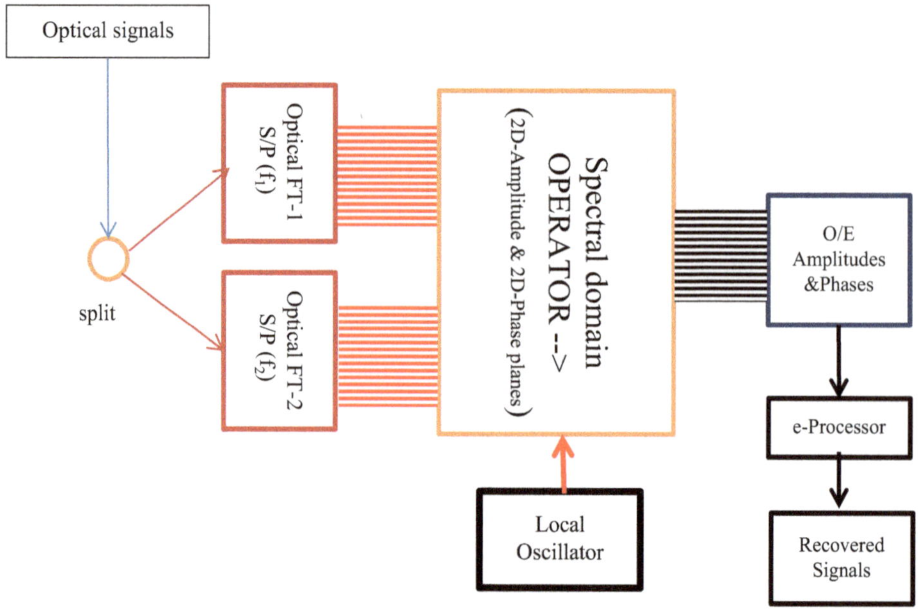

Figure 1.14. Generic model of spectral domain PSP. S/P = serial-to-parallel; FT = Fourier transform; O/E = optical to electrical converter (photodetector bank); 2D = two-dimensional; e- = electrical/electronic.

outputs of the OCNN-1 and OCNN-2 are compared (e.g. via an optical NAND), thus this is more or less an operation of differential processing. Such differential processing can be highly useful for differentially coded sequences as in differential coding or forward error coded (FEC) sequences.

In the case of spectral domain PSP (SPSP), we have as follows. The input signal sequence is split into two and then Fourier transformed into two spectral regions of fundamental frequencies f_1 and f_2, in parallel form (see figure 1.14).

These spectral FT optical sections are fed into an optical operator which can be bispectrum processors whose outputs are two 2D planes of amplitude and phase via a coherent detection sub-system. OCNN in the spectral domain can be used in the SPSP with a multiplication in the optical domain to represent the equivalence of time-domain auto- or cross-correlation operations if the processing is more complex.

In summary, we can see that both time domain and spectral domain PSPs can be used in receiving sub-systems or pre-processing optical transmitters, or in optical switching or routing. The applications of these generic models are given in this proposal. The demonstration of a PSP is to use a simple model of time-domain PSP without an optical feedback model.

We can compare the performance and processing parameters of PSP in contrast to a typical e-DSP based optical receiver as tabulated in table 1.1.

Table 1.1. Comparing a PSP and an e-DSP based optical receiver.

Parameters	e-DSP optical receiver	PSP	SPSP	Remarks
Sampling	Limited by e-sampler <200 GSa s^{-1}	Limited by delay optical waveguide >0.11 ps and optical phase modulation of optical switching in the sampler.	Limited by optical delay in Opt FT (MZDI).	PSP sampling rate ~10–20 DSP.
Noises	ENOB < 4.5 for 100–200 GSa s^{-1}	Not limited by ENOB but optical noise if optical OA is used.	Same as PSP.	OA = optical amplifier; PSP noise performance is much superior.
Processing speed	Limited by electronic processor (FPGA or ASIC) and memory length	Much faster than e-DSP but limited by memory—max memory length is ONE. Longer if OA used but oscillation is a problem.	Not limited by memory length and possible processing of long sequences by using the Wiener–Khinchin theorem and processing in the frequency domain of recovered elect sequence.	The e-DSP memory length is long and >1 million bit length.
Accuracy	Limited by the number of bit resolution, max 8-bt at 200 GSa s^{-1}.	Limited by optical samples by samplers and optical losses.	Limited by optical FT or MZDI, hence the number of optical filters.	PSP affected by optical responses of optical samplers. e-DSP not much due to logical/digital levels.
Complexity	e-DSP possible processing complex signals, but processing time is long due to sequential process, particularly in correlation.	Simpler with addition in the optical domain for multiplication in the time domain and optical multiplication for correlation.	Addition for correlation operation in the spectral domain.	PSP much simpler and low noise—but limited by the response time as the PSP operation is an analog process.
Digital/analog processing	Digital	Analog	Analog	Limited by response time if analog operation.
Power consumption	High (>20 W for 200 GSa s^{-1})	<100 mW for 100 GBd rate	<100 mW	High for e-DSP.
Cost	High	Low if not limited by integrated photonic integrated circuit technology and memory.	Same as PSP.	PSP superior advantage if limited factors are not critical factors.

1.6 Modulation techniques for ultra-broadband

1.6.1 Modulation formats and optical signal generation

Modulation is a process of facilitating the transfer of information over a medium. In optical communications the process of converting information so that it can be successfully sent through the optical fiber is call optical modulation.

There are three basic types of digital modulation techniques. These are amplitude shift keying (ASK), frequency shift keying (FSK) and phase shift keying (PSK) in which the parameter is the carrier whose amplitude, frequency or phase is varied to represent the information which is to be sent. Digital modulation is a process of mapping such that the digital data of '1' and '0' or symbols of '1' and '0' that convert into some aspects of the carrier, such as the amplitude, the phase or both amplitude and phase, are then transmitted by the modulated carrier, which is lightwaves in the context of this book. The modulated and transmitted lightwave carrier is then remapped at the reception systems back to a near copy of the information data.

1.6.1.1 Binary level

Modulation is a process that facilitates the transport of information over the medium, in this book our medium is the optical guided fiber and associate photonic components. In digital communications there are three basic types of digital modulation techniques: ASK, PSK and FSK. All these techniques vary a parameter of a sinusoidal carrier to represent the information '1' and '0'.

In ASK the amplitude of the lightwave carrier, normally generated by a narrow linewidth laser source, is changed in response to the digital data and everything else is kept fixed. That is, bit '1' is transmitted by the lightwave carrier of a particular amplitude. To transmit '0' the amplitude is changed, keeping the frequency unchanged, as shown in figure 1.15. NRZ or RZ can be assigned depending on the occupation of the state '1' during the time length of a bit period. For RZ normally only half of the bit period is occupied by the digital data.

In addition to the NRZ and RZ formats, in optical communications the carrier can be suppressed under these formats so as to achieve non-return-to-zero carrier suppression (NRZ-CS) and return-to-zero carrier suppression (RZ-CS). This is normally generated by biasing the optical modulator in such a way that the carriers passing through the two parallel paths of an interferometric modulator are π-phase difference with each other. Thus the optical central carrier is suppressed and the two sidebands remain representing the modulated information in the spectral domain.

In PSK the phase of the lightwave carrier is changed to represent the information. The phase in this context is the shift of angle at which the sinusoidal carrier starts. To transmit a '0' the phase would be shifted by π and a '1' with no change of phase. The phase angle can be changed and take a value of a set of phases corresponding to the mapping of the symbols as shown in figure 1.16.

In FSK the frequency of the carrier represents the digital information. One particular frequency is assigned to a '1' and another frequency is assigned to the '0', as shown in figure 1.17. An FSK can be considered as continuous phase modulation, for example the continuous phase modulation minimum phase shift keying (MSK)

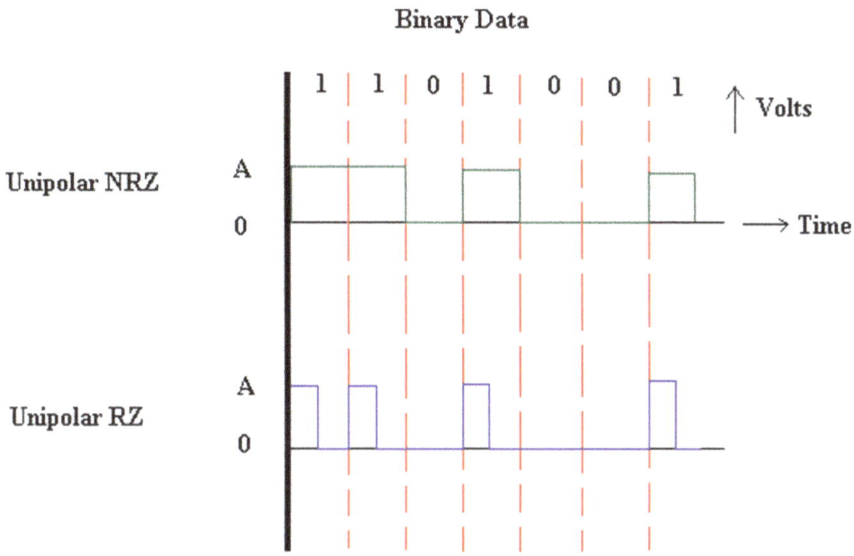

(a)

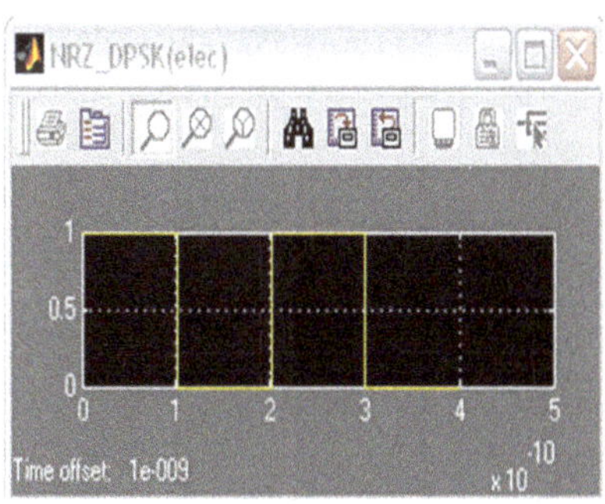

(b)

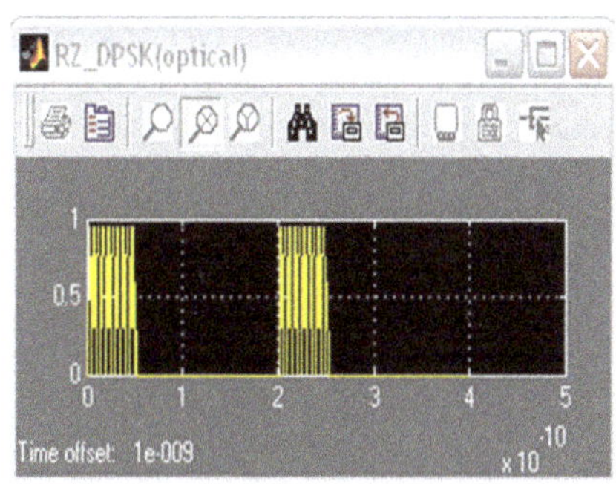

(c)

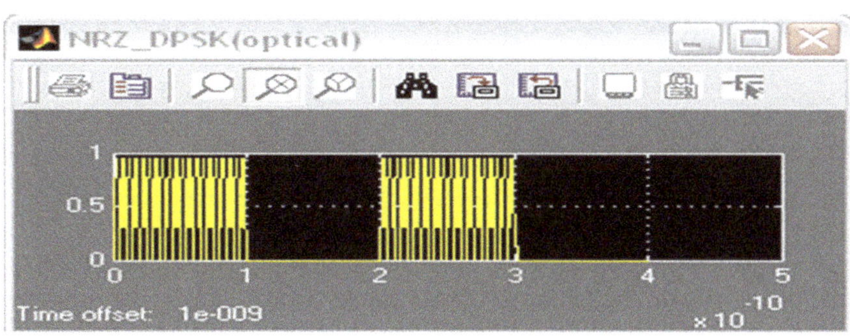

(d)

Figure 1.15. (a) Schematics of NRZ and RZ pulse amplitude modulated formats for a sequence of {1 0 1 0 1 0 1 0 1 0 1 0}. (b) Generated ASK signals with carrier (not to scale and high density area) data and carrier modulated. (c) NRZ and (d) NZ formats.

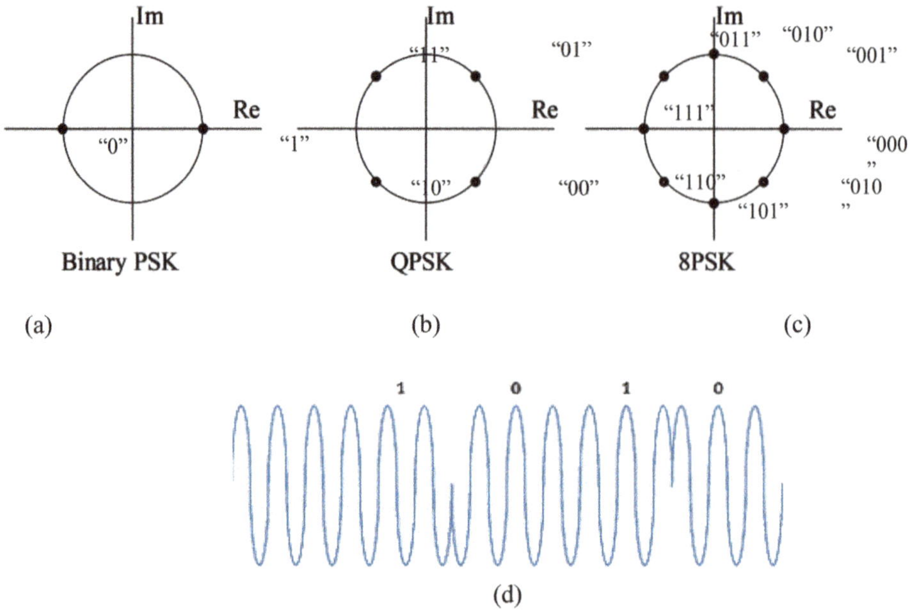

Figure 1.16. Signal space constellation of discrete phase modulation: (a) binary PSK, (b) quadrature binary PSK and (c) 8PSK. (d) Phase of the carrier under modulation with π phase shift of the BPSK at the edge of the pulse period.

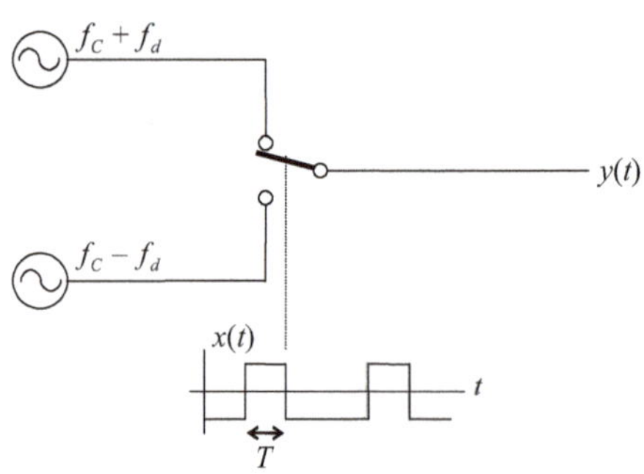

Figure 1.17. Schematic diagram of an FSK transmitter. f_d is the deviation frequency from the carrier frequency f_c.

whose frequency separation f_d *is* selected such that the signals carried by these frequencies are orthogonal.

ASK can be combined with PSK to create a hybrid modulation scheme, such as quadrature amplitude modulation (QAM) where both the phase and amplitude of the carrier are changed at the same time. The carriers are expected to follow a similar pattern to that of the DPSK in figure 1.17(d) but with different frequencies of the carrier under the envelope of the bit '0' and '1'. For MSK signals the carrier frequency is chirped up or down depending on the '0' or '1'. That is, the phase of the carrier is continuously varied during the bit period, and the carrier frequencies of the bits are such that there is an orthogonality of the carriers and the signal envelope.

1.6.1.2 Binary and multi-level

Additional degrees of freedom for detection can be used to effectively enhance the capacity due to effective equivalence of the multi-level and symbol rate, thence the detection of the received optical signals. A widely used and matured detection scheme for optical signals is direct detection, in which the optical power $P = [E]^2$, the square of a complex optical field amplitude. The photodetector (PD) would not be able to distinguish between a '0' or 'π' phase shift of the carrier lightwave embedded within the pulse. The carrier phase can only be extracted if and only if there is photonic processing to extract the phase at the front end of the receiver. Thus a + or – field complexity would be seen as identical in the PD. This ambiguity of the phase detection process would allow one to shape the optical spectra of optical signals to induce a modulation format more resilient to the distortion effects accumulated during the transmission process.

Formats making use of the tri-level can be illustrated in figure 1.18 and could be termed as pseudo-multi-level, tri-level, or polybinary signals. These tri-level signals can be represented in terms of the phase or frequency of the lightwave carrier.

The use of more than two symbols to encode a single bit of information is to increase the information bits carried by a symbol. However, the transmission is still at the symbol rate B_S. Under optical transmission, the -1 and $+1$ can be coded in terms of the variation of the phase of the carrier, as there is no negative intensity representation unless the field of the lightwaves is used. The tri-level uses $\{+|E|, -|E|, \text{and } 0\}$ (in field amplitude) or in optical domain the phase of the carrier can be coded as $\{+\pi/2, -\pi/2, \text{and } 0\}$ with its equivalent phase representation mapped to $\{0, |E|^2\}$ at the optical receiver, for example the duo-binary format that will be described in chapter 4. A phase difference of π and 0 between the three levels would minimize the pulse dispersion as they are propagating along the fiber due to the relative phase difference of π, hence destructive interference of any pulse spreading due to dispersion.

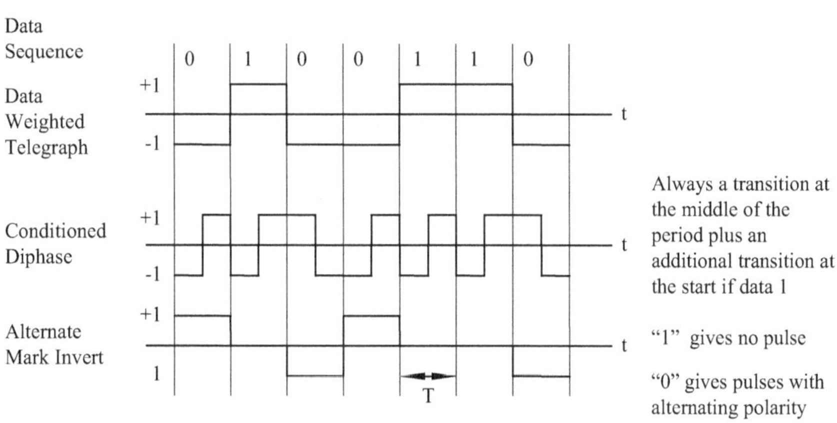

Figure 1.18. Illustration of pseudo-multi-level or polybinary baseband signals. Binary data sequence, weighted signals, diphase RZ and alternate mark inversion formats.

This tri-level must not be mixed up with the truly multi-level signaling in which log2M bits are encoded on N symbols, and then transmitted at a reduced rate B_s/log$_2N$. Both multi-level amplitude or amplitude-phase shift keying and DQPSK are multi-level optical modulation techniques.

1.6.1.3 In-phase, quadrature phase and amplitude modulation
Another form of modulation that would enhance the capacity of the transmission is the use of the orthogonal channels in which the information can be coded into the in-phase and quadrature in its polar or Cartesian coordinates, as shown in figure 1.19.

QPSK is the most commonly used in the differential and non-differential phase modulation in which the I and Q components are used extensively, due to its bit error rate (BER) and its corresponding energy per bit being very similar to that of a PSK and doubling the capacity of that of PSK. QPSK is an extension of the binary PSK signals but with the phase change of only $\pi/2$ instead of π. Mathematically the signal $s(t)$ can be expressed as

$$s(t) = A_c p_s(t)\cos\left(2\pi f_c t + i\frac{2\pi}{M}\right); \quad i = 1, \ldots, M, \tag{1.1}$$

where $p_s(t)$ is the pulse shaping of the data, M is the quantized level or the total number of phase states of the modulation and i is the phase modulation index. The QPSK can be combined with ASK to generate QAM where the phase and amplitude can be used to map a symbol of data information into one of the points on the signal space.

1.6.1.4 External optical modulation
External modulation is the essential technique for modulating the lightwaves so that the linewidth preserves its narrowness and only the sidebands of the modulation scheme dominate the spectral property of the generated passband characteristics. Figures 1.20 and 1.21 show the typical structure of optical transmitters for the generation of NRZ and RZ optical signals. The laser is always switched on and its lightwaves are modulated via the electro-optic (E/O) modulator using the principles of interferometric constructive and destructive interference to represent the ON and OFF states of the lightwaves. The RZ can then be generated similarly but with an additional optical modulator that would generate periodic optical pulses whose

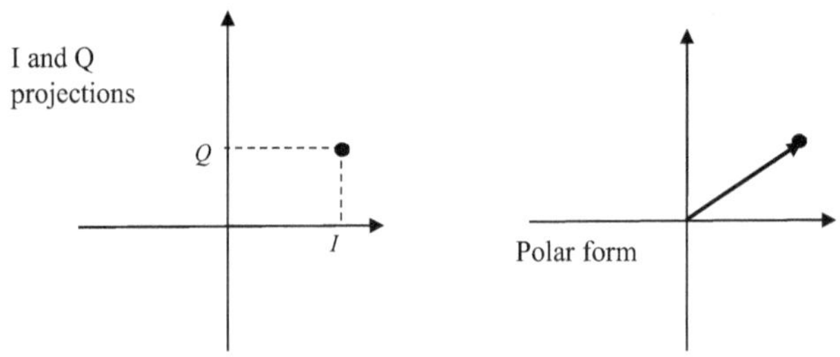

Figure 1.19. Signal vectors plotted in signal space: (a) Cartesian coordinate and (b) polar form.

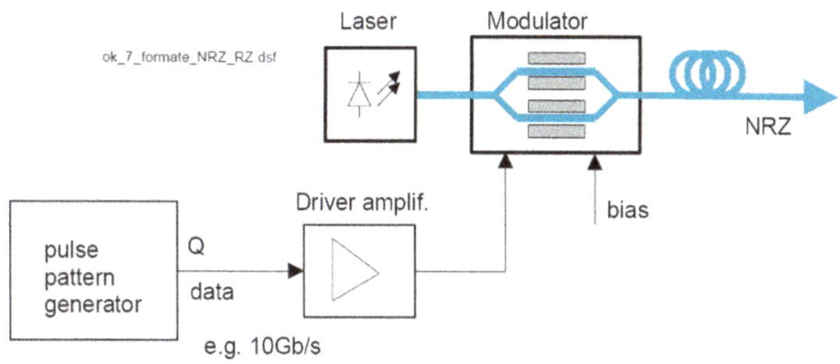

Figure 1.20. Generation of optical signals of format NRZ using an external modulator.

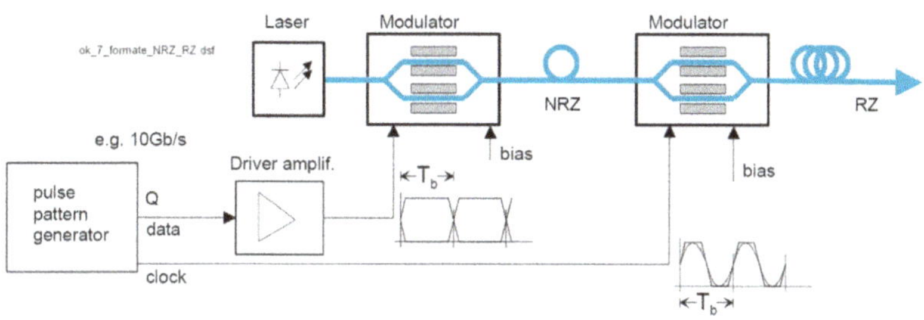

Figure 1.21. Generation of optical signals of format RZ.

width is half that of the bit period. The phase and frequency modulation can also be generated using these E/O modulators by biasing conditions and control of the amplitudes of the electrical pulses. These optical transmitters are described in [28]. We note here that the fiber that connects the two modulators of figure 1.19 must be of polarization maintaining (PM) type. Otherwise there would be polarization fluctuation and hence reduction of the coupling of the lightwave power one to the other.

The laser source would normally be a narrow linewidth laser that is turned on at all times to preserve its narrowness characteristics. The lightwaves are generated and coupled to the optical modulator via the pigtails of both devices. The modulator would be driven by a data pulse sequence output of a bit pattern generator condition to the appropriate driving level required by the V_π and the phase variation of the carrier if phase modulation is needed. When a modulation format is necessary then an electronic pre-coder is required to code the serial sequence to appropriate coding. The pre-coder can be a differential coding, multi-level coding or inverse fast Fourier transform to generate orthogonal data sub-channels in the case of the OFDM.

The pulses of all modulation formats can take the form of NRZ or RZ. For the RZ format an additional optical modulator is required to generate or condition the '1' NRZ to RZ, as shown in figure 1.21. The second modulator can exchange its position with that of the other modulator without affecting the generation of the modulation formats. This second modulator is usually called the pulse carver. It is noted that if the RZ modulator is biased such that the phase difference at the biasing condition is π phase difference then we would have a carrier suppression of the carriers at the central location of the generated spectra but the sidebands of the optical signals. The bandwidth of the modulator determines the rise time and fall time of the edges of the pulse sequence shown in figures 1.20 and 1.21. The details of the optical transmitters for different modulation formats are given in [29].

1.6.1.5 Advanced modulation formats

The above-described problems facing contemporary optical fiber communications can be effectively overcome by utilizing spectrally efficient transmission schemes via the implementation of advanced modulation formats. A number of modulation formats have recently been reported as alternatives for the OOK format, including return-to-zero (RZ) pulses in OOK/ASK systems [19–21], differential phase shift keying (DPSK)[1,2] [19–23] and, more recently, minimum-shift keying (MSK) [30–34]. These formats are adopted into photonic communications from the knowledge of wire-line and wireless communications.

DPSK has received much attention for the last few years, particularly when it is combined with RZ pulses. The main advantages of RZ DPSK are: (i) a 3 dB improvement in the receiver sensitivity over the OOK format by using an optical balanced receiver [28] and (ii) high resilience to fiber nonlinearities [6, 19, 29, 35] such as intra-channel SPM and inter-channel XPM. Several experimental demonstrations of DPSK long-haul DWDM transmission systems for 10 Gb s^{-1}, 40 Gb s^{-1} and higher bit rates have been reported recently[1] [22, 36–39]. However, there are few practical experiments addressing the performance of cost-effective 40 Gb s^{-1} DPSK–10 Gb s^{-1} OOK hybrid systems for gradually upgrading the existing installed SSMF transmission infrastructure [40]. In addition, the performance of 40 Gb s^{-1} DPSK for use in this hybrid transmission scheme has not been thoroughly studied. Therefore, one of the main contributions of this research is to prove the feasibility of overlaying 40 Gb s^{-1} DPSK channels on the existing 10 Gb s^{-1} network infrastructure for implementing hybrid systems.

The MSK format offers a spectrally efficient modulation scheme compared to the DPSK counterpart at the same bit rate. As a subset of continuous phase frequency shift keying (CPFSK), MSK possesses spectrally efficient attributes of the CPFSK family. The frequency deviation of MSK is equal to a quarter of the bit rate and this frequency deviation is also the minimum spacing to maintain the orthogonality between two shifted frequencies of FSK. On the other hand, MSK can also be considered as a particular case of offset differential quadrature phase shift keying (ODQPSK) [41–44], which enables MSK to be represented by I and Q components on the signal constellation. The advantageous characteristics of the optical MSK format can be summarized as follows. (i) A compact spectrum, which is of particular interest for spectrally efficient and high speed transmission systems—this also provides robustness to tight optical filtering. (ii) High suppression of spectral side lobes in the optical power spectrum compared to DPSK. The roll-off factor follows f^{-4} rather than f^{-2} as in the case of DPSK. This also reduces the effects of inter-channel cross-talk. (iii) No high power spectral spikes in the power spectrum, thus reducing fiber nonlinear effects compared to OOK. (iv) As a subset of either CPFSK, MSK can be detected either incoherently based on the phase or the frequency of the lightwave carrier, or coherently based on the popular *I–Q* detection structure. (v) The constant envelop property, which eases the measure of the average optical power.

Several studies have been conducted recently on the generation and direct detection of externally modulated optical MSK signals[3,5] [24]. However, there are few studies investigating the performance of the externally modulated MSK format for digital photonic transmission systems, particularly at high bit rates such as 40 to 100 Gb s^{-1} over four lanes[3] [45, 46]. Furthermore, if MSK can be combined with a multi-level modulation scheme, the transmission baud rate would be reduced in addition to the spectral efficiency of the MSK formats. This is of great interest for long-distance and metropolitan optical networks and, thus, provides the main motivation for proposing the dual-level MSK modulation format in this study. In addition, the potential of optical dual-level MSK format transmission has yet to be explored. Therefore, another main contribution of this research is to provide comprehensive studies on the performance of MSK and dual-level MSK modulation formats for long-distance and metropolitan optical transmission systems.

1.6.1.6 Orthogonality spectral modulation: OFDM and DMT

Orthogonal frequency division multiplexing (OFDM) is an attractive modulation format that has recently received lots of attention in practical and R&D fiber-optic communities [47]. The main advantage of optical OFDM is that it can cope with a virtually unlimited amount of inter-symbol interference (ISI). This is possible as the whole band is divided into several subcarriers, and modulation formats can be assigned to each carrier of much narrower bandwidth. Thus the dispersion becomes negligible. In high speed optical transmission systems, ISI caused, for example, by chromatic dispersion (CD) and polarization mode dispersion (PMD), are serious in long and short distance systems whose bit rate is higher than 56–100 GBd, in particular at these ultra-high rates and in multi-level modulation in the optical domain. In this sub-section some of the basics of an optical OFDM system are given. Furthermore, high speed transmission experiments up to 122 Gbps are briefly presented.

In principle, the basic concept behind OFDM is the division of a high bit rate data stream into several low bit rate orthogonal subcarrier streams. In general, the subcarriers are generated in the electronic digital domain, therefore these systems typically consist of many subcarriers (typically more than 64 and should be an exponential number with 2 as the base). In these systems, channel estimation is realized by periodically inserting training symbols. In fiber-optic transmission systems, OFDM systems where the subcarriers are generated in the optical domain are also created. These systems are sometimes referred to as coherent WDM systems [47]. Coherent WDM systems typically have few subcarriers and do not use training symbols, but rely on blind channel estimation instead.

In optical OFDM systems, the front end of the transmitter consists of an optical modulator, where the OFDM signal is unconverted to the frequency of an optical carrier. The front end at the receiver consists of either a coherent (Co) or a direct detection (DD) scheme. DD-OFDM is realized by sending the optical carrier along with the OFDM band so that direct detection with a single photodiode can be used at the receiver. In a Co-OFDM system, the optical carrier is suppressed at the transmitter and a local oscillator (LO) and an optical hybrid is required such as that shown in figure 1.2. The superior performance of Co-OFDM with respect to optical signal-to-noise ratio (OSNR) requirements, PMD tolerance and spectral efficiency makes it an excellent candidate for long-distance transmission systems, whereas DD-OFDM, which requires fewer components at the receiver than Co-OFDM, is more suitable for cost-effective short reach applications. The spectra of both these two OFD schemes are shown in figure 1.22. The DSP processing of DD-OFDM is much simpler as only the real part of the OFDM signals is required to be generated. The alternative name for DD-OFDM is discrete multi-tone (DMT). For DMT the total capacity can be increased using the bandwidth extending to 6 dB rather than only 3 dB. For the subcarriers in the 3 dB to 6 dB range, the subcarrier can be modulated at lower order of modulation formats so that the sensitivity of the receiver can be increased to compensate for the lower sensitivity by the 3 dB lower band, as shown in figure 1.23. OFDM and DMT are described in detail in chapter 3.

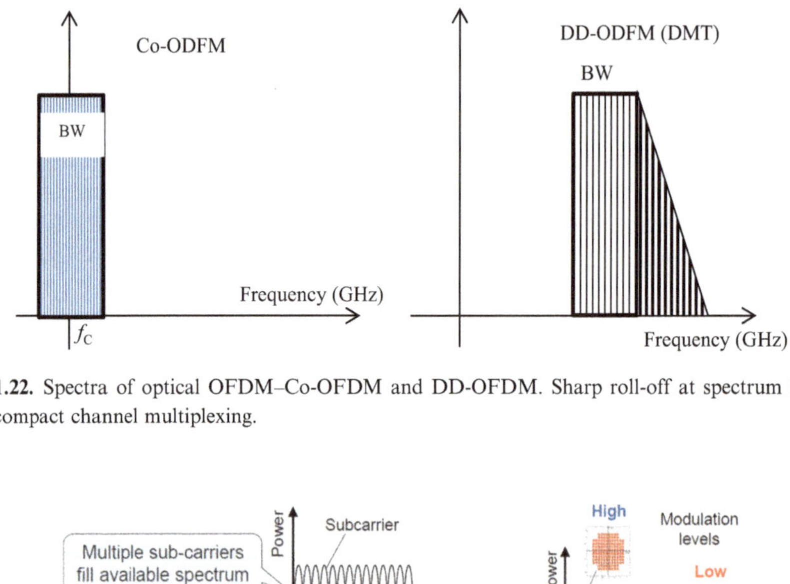

Figure 1.22. Spectra of optical OFDM–Co-OFDM and DD-OFDM. Sharp roll-off at spectrum boundary thence compact channel multiplexing.

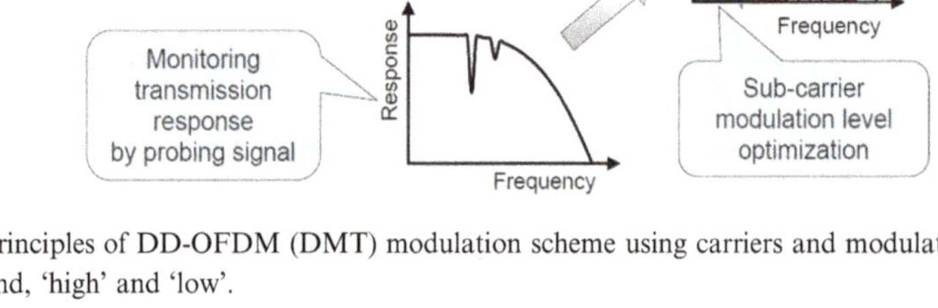

Figure 1.23. Principles of DD-OFDM (DMT) modulation scheme using carriers and modulation formats in the 0–6 dB band, ‘high’ and ‘low’.

1.7 Opto-electronic reception and processing

The reception sub-system is an important part of an optical communication link. Its sensitivity is critical for determining the transmission speed and reach distance. The O/E is the opto-electronic device (PD) which converts the optical energy into electrical signals. However, if a local oscillator laser is combined with the signal optical field and injected into such a PD then the optical fields of the transmitted signals and the local laser are mixed PD to give a product in the RF spectral region, the direct current (dc) component due to the residual power of the local oscillator and that of the signals which are positioned at the dc. These dc components are normally filtered and only the RF component remains to be detected, whose phase and amplitude reflect those of the signals multiplied by the amplitude of the LO. Thus it can be considered as an amplification of the signal field by the amplitude of the LO with its phase. Therefore, phase dependent modulation is effectively detected by this coherent mixing and detection. This section gives a brief introduction to these reception sub-systems.

1.7.1 Incoherent optical receivers

The modulation formats studied in this research, optical DPSK and MSK-based formats, can be demodulated incoherently using an optical balanced receiver which employs a Mach–Zehnder delay interferometer (MZDI). In the case of the optical DPSK format, MZDI is used to detect differentially coded phase information between every two consecutive symbols [6, 29, 30, 48]. This detection is carried out in the photonic domain as it is beyond the speed of the electrical domain, in particular at very high bit rates of 40 Gbps or above. The MZDI balanced receiver is also used for incoherent detection of optical MSK signals[3,5] [24] by also detecting the differential phase of MSK-modulated optical pulses. However, using the MZDI-based detection scheme, it is found that optical MSK provides a slight improvement for the CD tolerance over DPSK and OOK counterparts [24, 25].

As a subset of the CPFSK family, MSK-modulated lightwaves can also be incoherently detected based on the principles of optical frequency discrimination. Thus, an optical frequency discrimination receiver (OFDR) employing dual narrowband optical filters and an optical delay line (ODL) is proposed in this research. This receiver scheme effectively mitigates CD-induced ISI effects and enables breakthrough CD tolerances for optical MSK transmission, as reported in [25, 26]. In addition, the feasibility of this novel receiver is based on recent advances in the design of optical filters, in particular the micro-ring resonator filters. Such optical filters have very narrow bandwidths, e.g. less than 2 GHz (3 dB bandwidth), and they have been realized commercially by Little Optics [49, 50]. This research thus provides a comprehensive study of this OFDR scheme, from the operational principles to the analysis of the receiver design, and on to the performance of OFDR-based MSK optical transmission systems.

1.7.2 DSP—coherent optical receivers

Coherent detection and transmission techniques were extensively exploited in the mid-1980s to extend the repeaterless distance a further 20–40 km of SSMF with an expected improvement of the receiver sensitivity of 10–20 dB depending on the modulation format and receiver structure using phase or polarization diversity.

In general, a coherent receiver would operate on the beating of the received optical signals and that of the field of a local laser oscillator. The beating optical signals are then detected by the photodetector (PD) with the phase of the carrier preserved that permits the detection of the phase of the carriers. Hence the phase modulation and continuous phase or frequency modulation signals can be processed in the electronic domain. With the advancement of digital electronic processors, the processing of the received signals either in the IF or baseband of the heterodyne and homodyne detection, respectively, can be processed to determine the phase of the modulated and transmitted signals. Coherent receivers for different modulation formats are described at appropriate sections of the chapters of the book.

The three types of coherent receivers, namely homodyne, heterodyne and intradyne detection techniques, are possible and dependent on the frequency difference of zero, the intermediate frequency greater or smaller than the passband of the signals between the local oscillator laser and that of the signal carrier. With modern advanced optically amplified fiber communications, broadband ASE noise always exists and under coherent detection the beating between the local laser source and ASE dominate the electronic noise of the receiver at the front end. These noise considerations are described in [29].

The advanced aspects of coherent reception systems are currently being pushed to a new generation by incorporating the digital signal processing (DSP) sub-systems and naturally the ultra-high sampling rate ADC and DAC permitting the bit rate per channel reaching 100 Gb s^{-1} and higher by polarization division multiplexing (PDM) QPSK or an MQAM modulation scheme with M of 16 or 32 or even higher. The DSP in association with advanced algorithms has overcome a number of difficulties faced by 'analog' coherent reception techniques extensively reported in the 1980s, such as frequency matching of the local oscillator and that of the signal carrier; the clock recovery and generation of sampling for re-timing; and the compensation of dispersion due to CD and PMD impairments and nonlinear distortions. These DSP techniques are focused on in chapters 3 and 4, and chapter 5 covers the optical hardware and digitalization circuitry for such processing systems.

1.7.3 e-DSP electronic equalization

Electronic equalizers have recently become one of the most promising solutions for future high performance optical transmission systems. The technological development of Si–Ge has enhanced electron mobility and hence produced shorter rise and fall times of the pulse propagation, thus increasing the processing speed. The sampling rate can now reach several giga-samples/sec (GSa s^{-1}). This enables the processing of a 10 Gbps bit rate of the data channel without any difficulty

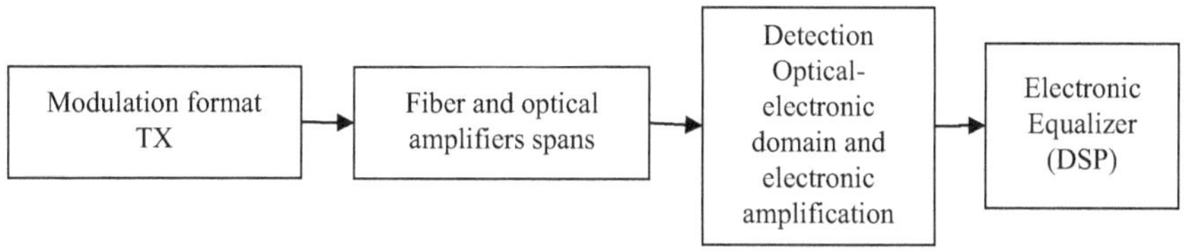

Figure 1.24. Schematic diagram of the location of the electronic equalizer at the receiver of an advanced optical communications system. DSP = digital signal processor.

(figure 1.24). Hence it is very probable that electronic processing and equalization can be implemented in real systems in the very near future.

The channel is single-mode optical fibers dispersive in the negative or positive factors and with some residual dispersion. Thus the distortion of the pulse is purely phase distortion prior to the detection by the PD, which follows the square law rule for the direct detection case. On the other hand, for coherent detection, the beating between the local oscillator and the signal in the PD would lead to the preservation of the phase, and one could be considered a pure phase distortion. Again, in the case of direct detection after the square law detection, the phase distortion is then transferred to the amplitude distortion. In order to conduct the equalization process it is important for us to know the impulse and step responses of the fiber channels $h_F(t)$ and $s_F(t)$. We will then give the fundamental aspects of equalization using feed forward equalization (FFE), decision feedback equalization with maximum mean square error (MMSE) or maximum likelihood sequence estimation (MLSE) of the Viterbi algorithm.

1.7.3.1 Feed forward equalizer

The FFE is the linear equalizer which has been the most widely studied, and a transversal filter structure would offer a linear processing of the signal prior to the decision. The structure of an FFE transversal filter consists of a cascade delay of the input sample and at each delay the signal is tapped and multiplied with a coefficient. These tapped signals, whose delay tap time is the bit period, are then summed to give the output sample. The coefficients of the transversal filter must take values that matched the channel so the convolution of the channel impulse response and that of the filter result in unity, so as to achieve a complete equalized pulse sequence at the output. Figures 1.25(a) and (b) show the linear equalization scheme that uses either feed forward or feedback equalization, the difference between these two schemes is the tapped signals either at the output of the transversal filter or at the output of the feedback that minimize the input sequence.

On the other hand a decision feedback equalizer (DFE) differs from that of the linear equalizer with a decision detector that would determine the signal amplitude required for feedback to the difference error at the input, as shown in figure 1.25.

1.7.3.2 Decision feedback equalization

The FFE method is based on the use of a linear filter with adjustable coefficients. The equalization method that exploits the use of previous detected symbols to

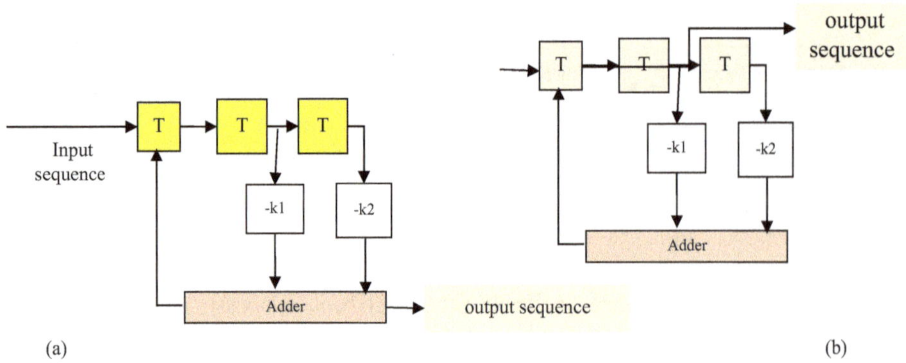

Figure 1.25. Linear (a) feed forward equalization (transversal equalization) and (b) feedback equalization scheme. $z^{-1} = T$ indicates the delay unit of sampling in the frequency domain or discrete domain.

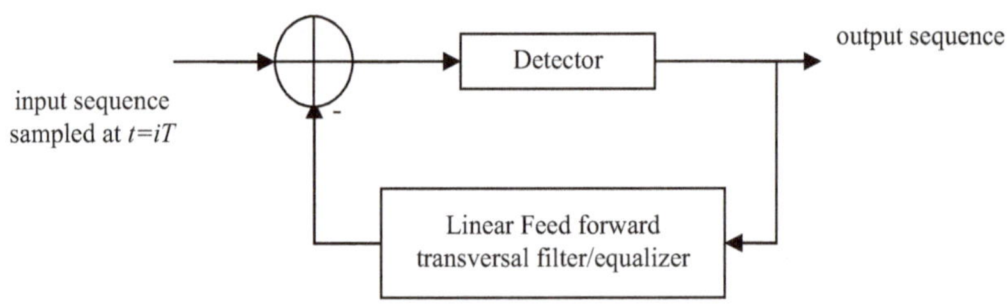

Figure 1.26. Schematic diagram of a receiver using nonlinear equalization by decision directed cancellation of ISI.

suppress the ISI in the present symbol being detected, is termed decision feedback equalization (DFE).

The DFE is illustrated in figure 1.26. It consists of m coefficients and m delay taps. Each tap spacing (delay in time domain) equals the bit duration T (or sampling time which may be much smaller). From figure 1.25(b) we can see that the received signal sequence goes through the forward filter first. After making decisions on previously detected symbols, the feedback filter provides information from the previously detected symbols for present estimating.

1.7.3.3 Minimum mean square error equalization

Consider next the general case where the linear equalizer is adjusted to minimize the mean square error due to both ISI and noise. This is called the minimum mean square error (MMSE) equalization.

1.7.3.3.1 Placement of equalizers

Linear and nonlinear equalization are possible as they are well known in the field of signal processing. The principles of equalization with equalizers placed at the transmitter, or the receiver, or shared between the transmitter and receiver, are critical for practical networks.

1.7.3.3.2 MLSE electronic equalizers

Amongst the electronic equalization techniques, maximum likelihood sequence estimation (MLSE) which can be implemented effectively with the Viterbi algorithm, has attracted considerable research interest [51–55]. However, most of the studies on MLSE equalizers have focused on either the ASK or DPSK formats [44, 45, 48, 56]. Apart from the recent paper [27], there has not been any study on the performance of MLSE equalizers for optical MSK transmission, in particular when OFDR is used as the detection scheme. The performance of OFDR-based MSK optical transmission systems is significantly enhanced with the incorporation of post-detection MLSE electronic equalizers as the ISI problems caused by either fiber dispersion impairments or tight optical filtering effects are effectively mitigated. Therefore, this decision process considered as an equalizer is included as a case study in chapter 4 of this book to comprehensively investigate the performance of MLSE equalizers for OFDR-based MSK optical transmission systems.

1.8 Organization of the book

The principal motivations of this book are to describe the processing of signals under the contemporary scenario of transmission technologies with ultra-high speed but short reach, as well as processing in the photonic domain, taking advantage of neural computing networks. Digital communications techniques in modern optical communications are now used under the scenario of cloud networking under extreme speeds, thus reaching the sampling rate limits of the electronic sampler in ADC or DAC. Short reach to medium transmission links have become much more important.

Furthermore, these techniques are essential for modules, integrated photonics, and processing in either the electronic or photonic domains for ultra-high capacity global data center inter-networking and particularly in the intra-DC networking and interconnecting between DCs, as well as ultra-high capacity access for distributed DC networking in 5G networks.

Chapter 2 provides a summary of the principles of optical transmission, in particular the modulation formats and receiving techniques to achieve ultra-wideband signals with band limited devices. The fundamental principles of digital communications employing either coherent or incoherent reception or associated modulation formats are described with a focus on the technological development and limitations in the optical domain. The enabling technologies, research results and demonstration in laboratory experimental platforms for the development of high performance and high capacity next-generation optical transmission systems will impose significant challenges to the engineering of optical transmission systems in the near future and techniques for next-generation data communications links.

The presentation of this book follows the progression of the digital communications and integration of modulation techniques in optical communications under self-coherent, direct detection and intensity modulation, and coherent reception techniques, as follows. The essential properties of optical fibers as the transmission medium for long-distance, metropolitan and access optical networks are introduced and given in the appendix. Both the geometrical and profile structures, as well as the propagation of modulation envelopes under the influence of interference due to the different propagation velocities, due to guided mode confinement of the fiber core and wavelength-dependent refractive index, are treated briefly to provide insight into the propagation of lightwaves and distortion of the modulated signals, so that the equalization and compensation in optical and DSP electronic domains can be developed in later chapters of the book.

The chapters are organized as follows:

In chapter 2, on modern DC networking (DCN):

- Evolution of DCN and traditional telecoms networks.
- Internet content transport for a high capacity demand society.
- Data transmission networks.
- Basic rates, capacity and server clusters.
- Inter-DC transport.
- Core and 'metro' transport.
- Latency and switching.

Chapter 3 presents optical transmission systems at ultra-high speed:

- Data optical communications at 400 G and higher.
- Terabits s^{-1} transport in DCN and DCI.
- Electronic processing in digital domain for ultra-speed transport.
- Modulation formats and capacity.

Chapter 4, on optical switching for data center networking:

- Electronic and optical switching for data communications.
- Photonic switching: technique and technology.
- Photonic switching devices and solutions for DCN and DCI.

Chapter 5, on photonic signal processing (PSP):

- Why PSP for data communications: speed and power consumption in DCN and DCI and impact of electronic processing and switching.
- PSP techniques.
- Optical neural computing and PSP.
- Multi-channel processing in the optical domain.

References

[1] Kao C and Hockham G 1966 Dielectric-fibre surface waveguides for optical frequencies *Proc. IEEE* **113** 1151–8

[2] Kaminow I P and Li T 2002 *Optical Fiber Communications* vol 4B (Amsterdam: Elsevier)

[3] Lin C, Kogelnik H and Cohen L G 1980 Optical pulse equalization and low dispersion transmission in single-mode fibers in the 1.3–1.7 mm spectral region *Opt. Lett.* **5** 476–8

[4] Gnauck A H, Korotky S K, Kasper B L, Campbell J C, Talman J R, Veselka J J and McCormick A R 1986 Information bandwidth limited transmission at 8 Gb/s over 68.3 km of single mode optical fiber *Proc. of OFC'86 (Atlanta, GA)* paper PDP6

[5] Kogelnik H 1985 High-speed lightwave transmission in optical fibers *Science* **228** 1043–8

[6] Agrawal G P 2002 *Fiber-Optic Communication Systems* 3rd edn (New York: Wiley)
[7] Chraplyvy A R, Gnauck A H, Tkach R W and Derosier R M 1993 8 × 10 Gb/s transmission through 280 km of dispersion-managed fiber *IEEE Photonics Technol. Lett.* **5** 1233–5
[8] Kogelnik H 2000 High-capacity optical communications: personal recollections *IEEE J. Sel. Top. Quantum Electron.* **6** 1279–86
[9] Giles C R and Desurvire E 1991 Propagation of signal and noise in concatenated erbium-doped fiber amplifiers *IEEE J. Lightwave Technol.* **9** 147–54
[10] Becker P C, Olsson N A and Simpson J R 1999 *Erbium-Doped Fiber Amplifiers, Fundamentals and Technology* (San Diego, CA: Academic)
[11] Farries M C, Morkel P R, Laming R I, Birks T A, Payne D N and Tarbox E J 1989 Operation of erbium-doped fiber amplifiers and lasers pumped with frequency-doubled Nd: YAG lasers *IEEE J. Lightwave Technol.* **7** 1473–7
[12] Okoshi T 1982 Heterodyne and coherent optical fiber communications: recent progress *IEEE Trans. Microwave Theory Tech.* **82** 1138–49
[13] Okoshi T 1987 Recent advances in coherent optical fiber communication systems *IEEE J. Lightwave Technol.* **5** 44–52
[14] Salz J 1986 Modulation and detection for coherent lightwave communications *IEEE Commun. Mag.* **24** 38–49
[15] Okoshi T 1986 Ultimate performance of heterodyne/coherent optical fiber communications *IEEE J. Lightwave Technol.* **4** 1556–62
[16] Henry P S 1990 *Coherent Lightwave Communications* (New York: IEEE)
[17] Elrefaie A F, Wagner R E, Atlas D A and Daut A D 1988 Chromatic dispersion limitation in coherent lightwave systems *IEEE J. Lightwave Technol.* **6** 704–10
[18] Charlet G *et al* 2005 Comparison of system performance at 50, 62.5 and 100 GHz channel spacing over transoceanic distances at 40 Gbit/s channel rate using RZ-DPSK *Electron. Lett.* **41** 145–6
[19] Cho P S, Grigoryan V S, Godin Y A, Salamon A and Achiam Y 2003 Transmission of 25-Gb/s RZ-DQPSK signals with 25-GHz channel spacing over 1000 km of SMF-28 fiber *IEEE Photonics Technol. Lett.* **15** 473–5
[20] Chagnon M, Osman M, Poulin M, Latrasse C, Gagné J-F, Painchaud Y, Paquet C, Lessard S and Plant D 2018 Experimental study of 112 Gb/s short reach transmission employing PAM formats and SiP intensity modulator at 1.3 μm *Opt. Express* **22** 21018–36
[21] Olmedo M I, Zuo T, Jensen J B, Zhong Q, Xu X, Popov S and Monroy I T 2014 Multiband carrierless amplitude phase modulation for high capacity optical data links *J. Lightwave Technol.* **32** 798–804
[22] Kai Y, Nishihara M, Tanaka T, Takahara T, Li L, Tao Z, Liu B, Rasmussen J and Drenski T 2013 Experimental comparison of pulse amplitude modulation (PAM) and discrete multi-tone (DMT) for short-reach 400-Gbps data communication *Proc. Eur. Conf. Optical Commun Conf. (ECOC)* paper Th.1.F.3
[23] Tanaka T, Nishihara M, Takahara T, Yan W, Li L, Tao Z, Matsuda M, Takabayashi K and Rasmussen J 2014 Experimental demonstration of 448-Gbps+ DMT transmission over 30-km SMF *Proc. Optical Fiber Commun. Conf. (OFC)* paper M2I.5
[24] Schares L *et al* 2006 Terabus: terabit/second-class card-level optical interconnect technologies *IEEE J. Sel. Top. Quantum Electron.* **12** 1032–44
[25] Shen Y *et al* 2017 Deep learning with coherent nanophotonic circuits *Nat. Photonics* **11** 441–6

[26] Paquot Y 2012 Optoelectronic reservoir computing *Sci. Rep.* **2** 28

[27] Fjelde T *et al* 2000 Demonstration of 40 Gbit/s all-optical logic XOR in integrated SOA-based interferometric wavelength converter *Elect. Lett.* **36** 1863–4
Younis R M *et al* 2014 Fully integrated AND and OR optical logic gates *Photonics Technol. Lett.* **26** 1900–03

[28] Gnauck A H and Winzer P J 2005 Optical phase-shift-keyed transmission *IEEE J. Lightwave Technol.* **23** 115–30

[29] Binh L N 2008 *Digital Optical Communications* (Boca Raton, FL: CRC Press/Taylor and Francis)

[30] Mo J, Yi D, Wen Y, Takahashi S, Wang Y and Lu C 2005 Optical minimum-shift keying modulator for high spectral efficiency WDM systems *Proc. of ECOC'05* **4** 781–2

[31] Huynh T L, Sivahumaran T, Binh L N and Pang K K 2007 Narrowband frequency discrimination receiver for high dispersion tolerance optical MSK systems *Proc. of Coin-Acoft'07 (Melbourne, Australia)* paper TuA1–3

[32] Huynh T L, Sivahumaran T, Binh L N and Pang K K 2007 Sensitivity improvement with offset filtering in optical MSK narrowband frequency discrimination receiver *Proc. of Coin-Acoft'07 (Melbourne, Australia)* paper TuA1–5

[33] Sivahumaran T, Huynh T L, Pang K K and Binh L N 2007 Non-linear equalizers in narrowband filter receiver achieving 950 ps/nm residual dispersion tolerance for 40 Gb/s optical MSK transmission systems *Proc. of OFC'07 (Anaheim, CA)* paper OThK3

[34] Sakamoto T, Kawanishi T and Izutsu M 2005 Optical minimum-shift keying with external modulation scheme *Opt. Express* **13** 7741–7

[35] Lazaro J A, Idler W, Dischler R and Klekamp A 2004 BER depending tolerances of DPSK balanced receiver at 43 Gb/s *Proc. of IEEE/LEOS Workshop on Advanced Modulation Formats 2004* 15–6

[36] Kim H and Gnauck A H 2003 Experimental investigation of the performance limitation of DPSK systems due to nonlinear phase noise *IEEE Photonics Technol. Lett.* **15** 320–2

[37] Bhandare S, Sandel D, Abas A F, Milivojevic B, Hidayat A, Noe R, Guy M and Lapointe M 2004 2/spl times/40 Gbit/s RZ-DQPSK transmission with tunable chromatic dispersion compensation in 263 km fibre link *Electron. Lett.* **40** 821–2

[38] Mizuochi T, Ishida K, Kobayashi T, Abe J, Kinjo K, Motoshima K and Kasahara K 2003 A comparative study of DPSK and OOK WDM transmission over transoceanic distances and their performance degradations due to nonlinear phase noise *IEEE J. Lightwave Technol.* **21** 1933–43

[39] Binh L N and Huynh T L 2007 Phase-modulated hybrid 40 Gb/s and 10 Gb/s DPSK DWDM long-haul optical transmission *Proc. of OFC'07 (Anaheim, CA)* paper JWA94

[40] Ito T, Sekiya K and Ono T 2002 Study of 10 G/40 G hybrid ultralong haul transmission systems with reconfigurable OADM's for efficient wavelength usage *Proc. of ECOC'02 (Copenhagen, Denmark)* paper 1.1.4

[41] Iwashita K and Takachio N 1989 Experimental evaluation of chromatic dispersion distortion in optical CPFSK transmission systems *IEEE J. Lightwave Technol.* **7** 1484–7

[42] Proakis J G 2001 *Digital Communications* 4th edn (New York: McGraw-Hill)

[43] Proakis J G and Salehi M 2002 *Communication Systems Engineering* 2nd edn (Englewood, NJ: Prentice Hall), pp 522–24

[44] Pang K K 2005 *Digital Transmission* (Melbourne: Mi-Tec)

[45] Iwashita K and Matsumoto T 1987 Modulation and detection characteristics of optical continuous phase FSK transmission system *IEEE J. Lightwave Technol.* **5** 452–60
[46] Mo J, Wen Y J and Wang Y 2007 Performance evaluation of externally modulated optical minimum shift keyed data *Opt. Eng.* **46** 035001
[47] Shieh W and Djordjevic I 2009 *OFDM for Optical Communications* 1st edn (New York: Academic)
[48] Huynh T L, Binh L N and Pang K K 2006 Optical MSK long-haul transmission systems *Proc. of SPIE APOC'06* paper 6353–86, Thu9a
[49] Elrefaie A F and Wagner R E 1991 Chromatic dispersion limitations for FSK and DPSK systems with direct detection receivers *IEEE Photonics Technol. Lett.* **3** 71–3
[50] Little B E 2003 Advances in microring resonator *Integrated Photonics Research Conf. 2003* Invited paper
[51] Alic N, Papen G C, Saperstein R E, Milstein L B and Fainman Y 2005 Signal statistics and maximum likelihood sequence estimation in intensity modulated fiber optic links containing a single optical preamplifier *Opt. Express* **13** 4568–79
[52] Brent L *et al* 2006 Advanced ring resonator based PLCs *IEEE Lasers Electro-Opt. Soc.* 751–2
[53] Curri V, Gaudino R, Napoli A and Poggiolini P 2004 Electronic equalization for advanced modulation formats in dispersion-limited systems *IEEE Photonics Technol. Lett.* **16** 2556–8
[54] Agazzi O E, Hueda M R, Carrer H S and Crivelli D E 2005 Maximum-likelihood sequence estimation in dispersive optical channels *IEEE J. Lightwave Technol.* **23** 749–62
[55] Napoli A 2006 Limits of maximum-likelihood sequence estimation in chromatic dispersion limited systems *Proc. OFC'06 (Anaheim, CA)* paper JThB36
[56] Qi J, Mao B, Gonzalez N, Binh L N and Stojanovic N 2013 Generation of 28 GBaud and 32 GBaud PDM-Nyquist-QPSK by a DAC with 11.3 GHz analog bandwidth *Proc. OFC2013* (San Francisco, CA)

Chapter 2

Data center networking

This chapter briefly describes the general aspects of data center networking (DCN), including some discussion of the historical evolution of the networking of data centers, both in terms of inter- and intra-structures. The technological development of transmission is described, including pulse modulation formats and quadrature amplitude modulation (QAM), or frequency-domain coherent orthogonal frequency division multiplexing (Co-OFDM) and direct detection OFDM (DD-OFDM), for increasing the total capacity rate with band limited components. Interconnection technology is also briefly described.

2.1 The evolution of DCN and traditional telecoms networks

A data center is a pool of resources (computational, storage and networking) interconnected using a communications network [1, 2]. The DCN has a pivotal role in a data center, as it connects all of the data center resources together. DCNs need to be scalable and efficient to connect tens or even hundreds of thousands of servers to handle the growing demands of cloud computing [3, 4]. Currently, data centers are constrained by the interconnection network [5].

2.1.1 Types of data center networks

2.1.1.1 Three-tier DCN

The three-tier DCN architecture, shown in figure 2.1, follows a multi-rooted tree based network topology composed of three layers of network switches, namely the access, aggregate, and core layers [6]. The servers in the lowest layers are connected directly to one of the edge layer switches. The aggregate layer switches interconnect multiple access layer switches. All of the aggregate layer switches are connected to each other by core layer switches. Core layer switches are also responsible for connecting the data center to the Internet. The three-tier DCN is the common network architecture used in data centers [6]. However, the three-tier architecture is unable to handle the growing demands of cloud computing [7]. The higher layers of

doi:10.1088/978-0-7503-2292-8ch2

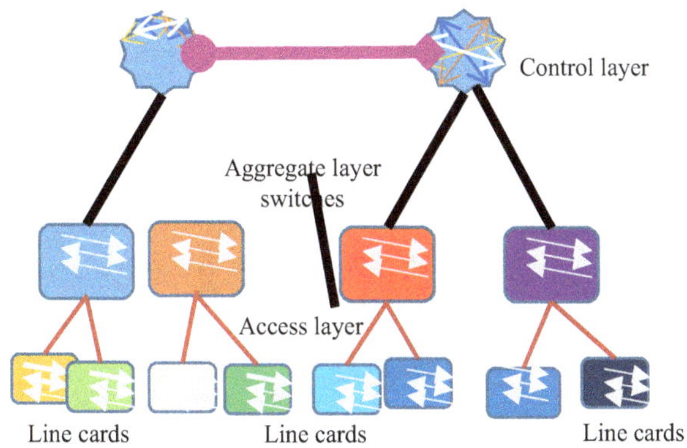

Figure 2.1. Three-tier structure of a DCN: core, aggregation and access layers. All connection paths are optical fibers of single mode type, the standard single mode fiber (SSMF).

the three-tier DCN are highly oversubscribed [3]. Moreover, scalability is another major issue in three-tier DCNs. Major problems faced by the three-tier architecture include scalability, fault tolerance, energy efficiency and cross-sectional bandwidth. The three-tier architecture uses enterprise-level network devices at the higher layers of topology that are very expensive and power hungry [5].

2.1.1.2 Fat tree DCN

The fat tree DCN architecture handles the over subscription and cross-section bandwidth problem facing the legacy of three-tier DCN architecture. The fat tree DCN employs commodity network switch based architecture using Clos topology [3]. The network elements in fat tree topology also follow the hierarchical organization of network switches in access, aggregate and core layers. However, the number of network switches is much larger than in the three-tier DCN. The architecture is composed of k pods, where each pod contains $(k/2)^2$ servers, $k/2$ access layer switches and $k/2$ aggregate layer switches in the topology. The core layers contain $(k/2)^2$ core switches where each of the core switches is connected to one aggregate layer switch in each of the pods. The fat tree topology offers a 1:1 over subscription ratio and full bisection bandwidth [3]. The fat tree architecture uses a customized addressing scheme and routing algorithm (figure 2.2). The scalability is one of the major issues in fat tree DCN architecture and the maximum number of pods is equal to the number of ports in each switch [7], as shown in figure 2.3.

2.1.1.3 DCell

DCell is a server-centric hybrid DCN architecture where one server is directly connected to many other servers [4]. A server in the DCell architecture is equipped with multiple network interface cards (NICs). The DCell follows a recursively built hierarchy of cells. A $cell_0$ is the basic unit of DCell topology, arranged in multiple levels, where a higher level cell contains multiple lower layer cells. The $cell_0$ is the building block of DCell topology, which contains n servers and one commodity network switch. The network switch is only used to connect the server within a $cell_0$. A $cell_1$ contains $k = n + 1$ $cell_0$ cells, and similarly a $cell_2$ contains $(k \times n + 1)$

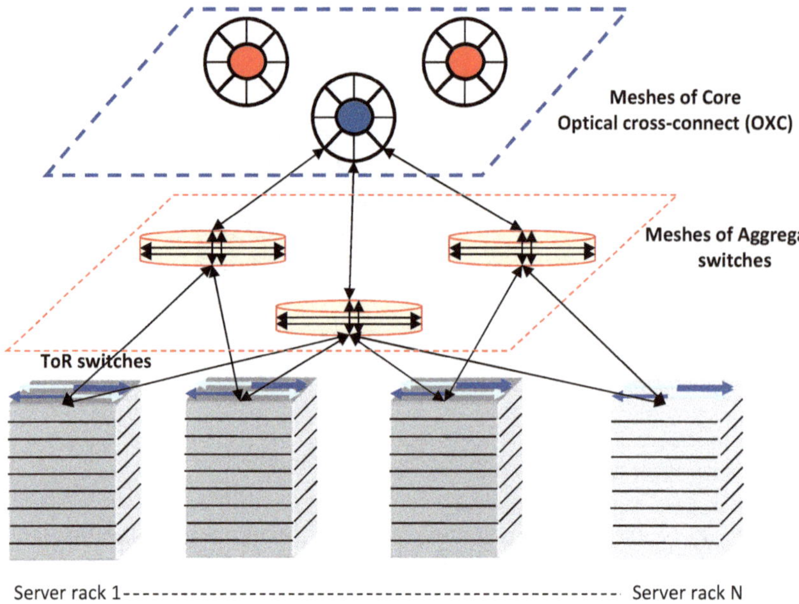

Figure 2.2. A multi-core aggregated switch three-tier DC with interconnection networking with clustered server racks. ToR = top of rack.

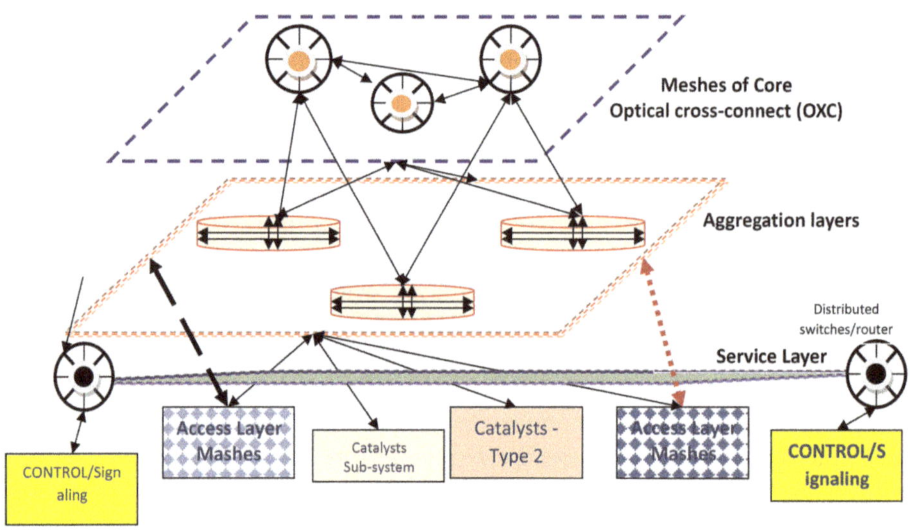

Figure 2.3. Detailed access layers and access interface devices with distinct service layers in both lower and upper levels of the aggregation and service layers, respectively, which follow the optical transport core networks. All lines are optical using standard SMF.

$cell_1$ cells. The DCell is a highly scalable architecture where a four-level DCell with only six servers in $cell_0$ can accommodate around 3.26 million servers (figure 2.4). In addition to its very high scalability, the DCell architecture presents a very high structural robustness [8]. However, cross-section bandwidth and network latency is a major issue in DCell DCN architecture [1].

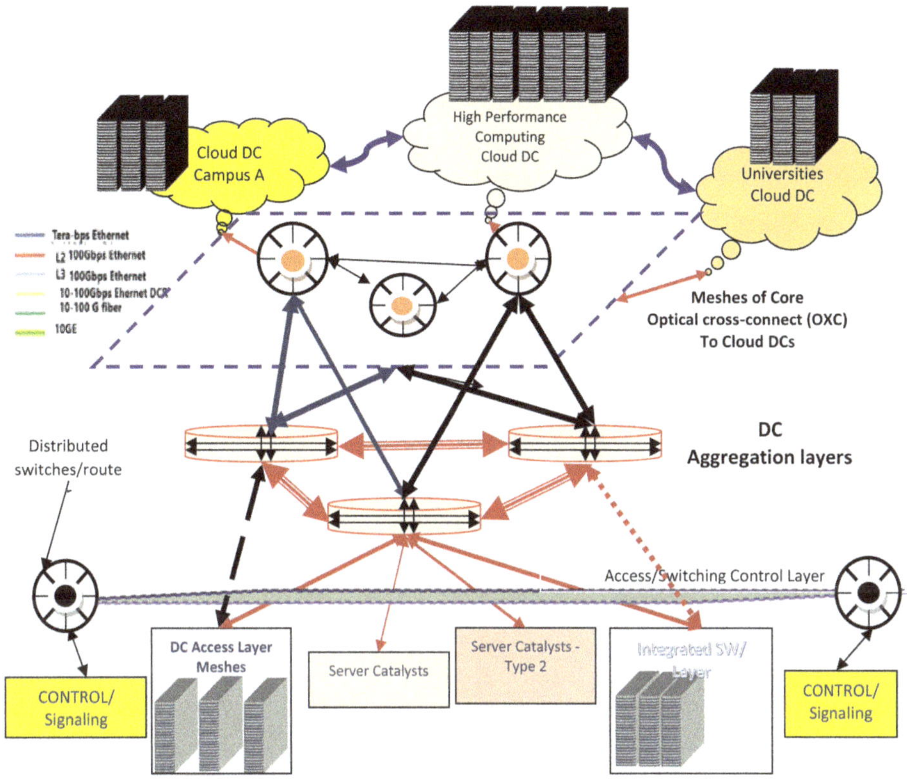

Figure 2.4. Arrangement of the structures of a DC with stack aggregation and access layers in enterprise DCNs, mostly for university academic campuses.

2.1.1.4 *Other types of DCN*

Other well-known DCNs include: BCube [9], Camcube [10], FiConn [11], Jelly fish [12] and Scafida [13] A qualitative discussion of different DCNs along with the benefits and drawbacks associated with each one are referred to in figure 2.5.

2.1.1.5 *Challenges*

Scalability is one of the foremost challenges for DCNs. With the advent of the cloud paradigm, data centers are required to scale up to hundreds of thousands of nodes. In addition to offering immense scalability, DCNs are also required to deliver high cross-section bandwidth. Current DCN architectures, such as three-tier DCN, offer poor cross-section bandwidth and possess a very high over subscription ratio near the roots [3]. Fat tree DCN architecture delivers a 1:1 over subscription ratio and high cross-section bandwidth, but it suffers from low scalability, limited to k as the total number of ports in a switch. DCell offers a high degree of scalability, but delivers very poor performance under a heavy network load and one-to-many traffic patterns (figure 2.6).

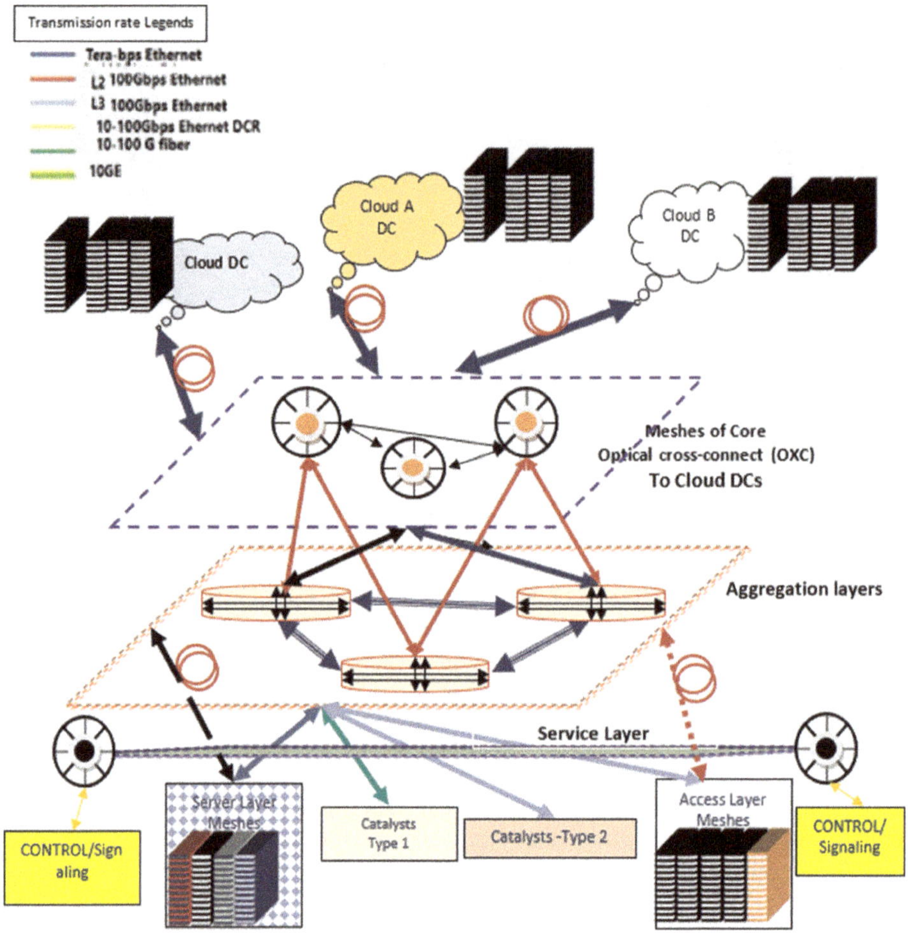

Figure 2.5. Generic model of data center networking in multi-terabits s^{-1} capacity. Model of transmission interconnection rates in DC networking and layers [6]. All connections are optical fibers—single mode and inline optical amplifiers.

2.1.2 Performance of DCNs

A quantitative analysis of the three-tier, fat tree and DCell architectures for performance comparison (based on throughput and latency) was performed for different network traffic patterns in [1]. The fat tree DCN delivers high throughput and low latency compared to three-tier and DCell. DCell suffers from very low throughput under high network load and one-to-many traffic patterns. One of the major reasons for DCell's low throughput is the very high over subscription ratio on the links that interconnect the highest level cells.

2.1.2.1 Structural robustness and connectivity of DCNs

The DCell exhibits very high robustness against random and targeted attacks and retains most of its nodes in the giant cluster after even 10% targeted failure [8], and

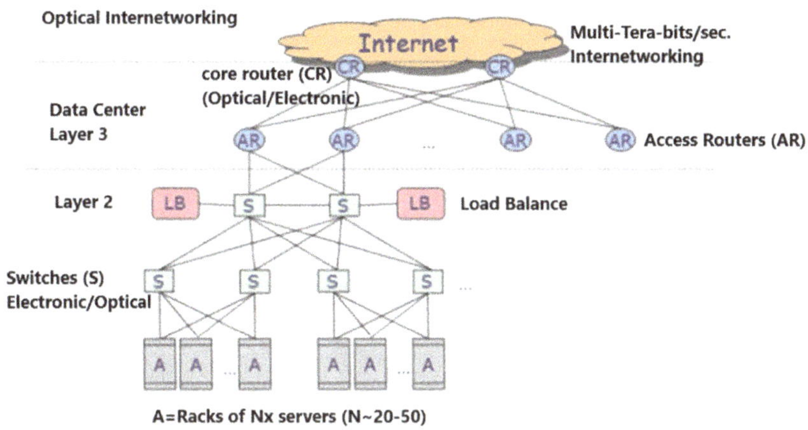

Figure 2.6. A three-layer optical Internet by DCN. All transmission and connection lines are optical using standard SMF.

multiple failures whether targeted or random, compared to the fat tree and three-tier DCNs [14]. One of the major reasons for the high robustness and connectivity of the DCell is its multiple connectivity to other nodes that is not found in fat tree or three-tier architectures

2.1.2.2 Energy efficiency of DCNs

The concerns about the energy needs and environmental impacts of data centers are intensifying. Energy efficiency is one of the major challenges of today's information and communications technology (ICT) sector. The networking portion of a data center is calculated to consume around 15% of overall cyber energy usage. Around 15.6 million MWh of electrical power, or a total of 15.6 trillion-watt (Tera-W) per hour was utilized solely by the network infrastructure within data centers worldwide. The energy consumption by the network infrastructure within a data center is expected to increase to around 50%. The IEEE 802.3 az standard was finalized in 2011, which makes use of an adaptive link rate technique for energy efficiency [15]. Moreover, the fat tree and DCell architectures use commodity network equipment that is inherently energy efficient. Workload consolidation is also used for energy efficiency, by consolidating the workload on a few devices to power-off or sleep for idle devices [16].

2.1.3 Communication interconnection speed

Due to the demand of ultra-high capacity transport for interconnections between data centers and distribution networks, the total transmission rate of information channels must be increased to the maximum possible level in order to minimize the numbers of optical fibers to be used for interconnections between servers and core switches within a data center, as depicted in figure 2.7.

Detailed access layers and access interface devices with distinct service layers in the lower and upper levels before to core layers. All lines are optical fibers of

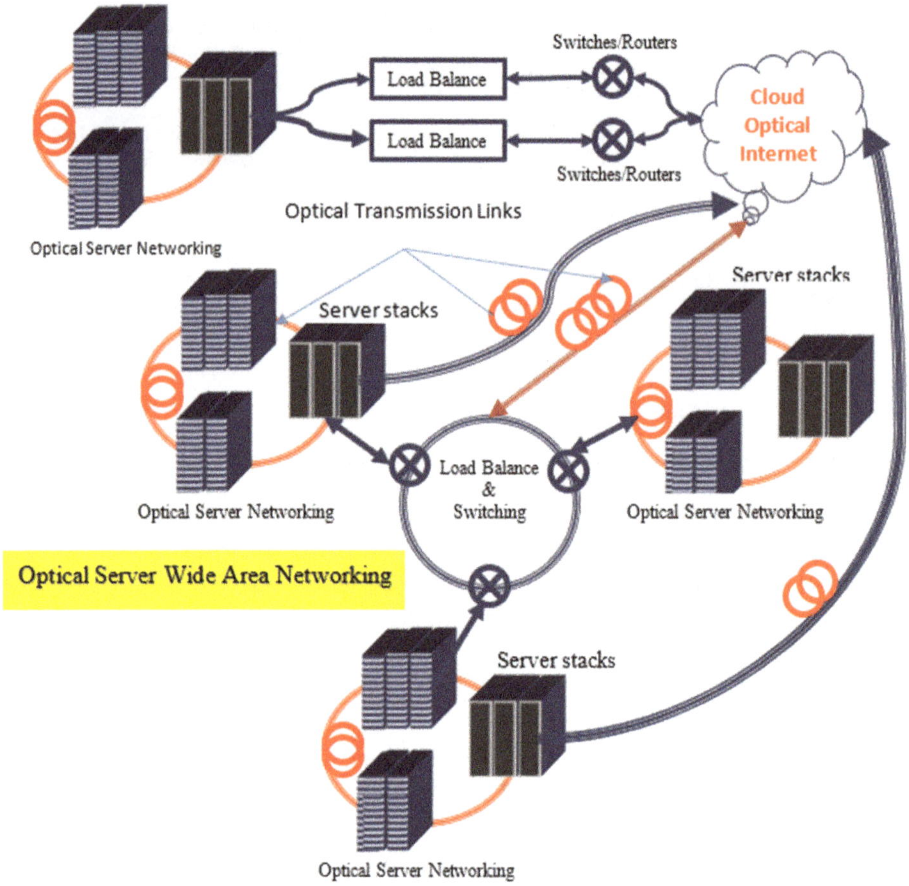

Figure 2.7. The contemporary global optical Internet in the distribution mode of DCs. LAN = local area network; WAN = wide area network; IP = internet protocol.

the standard SMF for long haul and metropolitan network transmission. All other connection lines such as access link or distribution networks may use SMF in different spectral regions, for example the O-band of 1310 nm or multimode or multicore fibers in order to minimize the costings.

Currently 400 G has been standardized and higher aggregate speeds over four wavelength channels or four lanes are being planned. It is one of the main aims of this book to describe and provide an understanding of the transmission technologies for such DC networking environments, for example DC networks in inter-university campuses.

2.2 Telecoms carriers and challenges from data center networking

Almost all carriers around the world, such as Korean, Telstra (Australia), Deutsche Telecom (Germany), Vodafone, Telefonica, etc, have been taking urgent action to flatten their traditional telecommunications networks to meet the demands of simplicity, minimum latency and high speed capacity transport. These are the principal superior properties expected from DCN communities such as Google, Facebook, Microsoft, etc. Both theoretic studies and demonstrations of 5G networks have been carried out to test some 5G features on live networks. The next sections of this chapter also demonstrate deployment tests of 5G in commercial environments [17]. We can expect that the flattening of traditional telecoms networks can match those of DCN with edge cloud distributed data centers of small or medium capacity, compared to concentric DCs.

A generic scenario of such 5G networking is depicted in figure 2.7, which gives an overview of 5G transport networks in a densely populated city, long-distance interconnected transport of telecoms carrier networks, DCs and user-concentrating environments, as well as machine-to-machine (M2M) and the Internet of everything (IoET).

2.2.1 Evolution to 5G transport DC based networks

We expect the following evolution of the transport networks (see figure 2.11):

- Massive evolution of transport networks from the IP-based architecture of 4G to a cloudified architecture for 5G, driven by changes in the ultra-high capacity and dynamic demand wireless and core network architectures of evolutionary traditional telecoms and cloud DC networking.
- Aggregation of DCs interconnected to metro core (MC) networks. The low latency and wideband services of 5G drives the down shifting of the backbone node to the metro node. Alternatively, the whole terrestrial or intercontinental network can be viewed as a metro transport network in which low cost low latency and ultra-high capacity must be met. These are the new challenges to be met by all, not only the DCN or telecoms carriers, as effectively all are 5G service providers (5GSP).
- As far as the RAN are to be densely distributed, metro D-RAN layers are formed and flattened, on which the cloud data units and centralized units can be aggregated.
- Cloud-RAN can enhance the gain of a base station in a 4G network by 10 dB based on test results from Japan, and more obviously in a 5G environment. This thus requires a unification of the common public radio interface (CPRI) and electrical CPRI.
- The optical line terminals (OLTs) will evolve into cloudified mobile engines (MCE) and receive or transmit to new 5G core nodes.
- 5G network architecture is trending towards common and dynamic broadband and metro private line networking, as well as multi-service transport.

The frequency bands for 5G, in particular $5G^{+}$ apart from the frequency band 2–6 GHz (frequency range 1 = FR 1), are located in the microwave and millimeter wave (mmW), centralized at 18.6 GHz and 28.6 GHz or 56.8 GHz and 90 GHz. The FR3 is assigned with 7 GHz bandwidth with its center located at 58.6 GHz. Data channels are to be delivered to mmW antennae, which are multiple-input multiple-output (MIMO) integrated horns, via an optical guided medium, and converted to

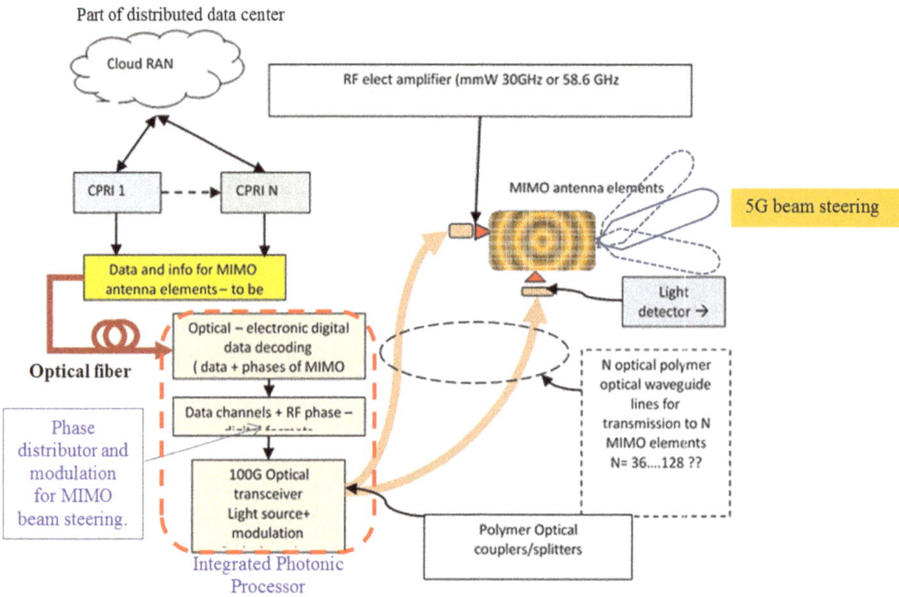

Figure 2.8. An integrated photonic processor in an OTTA.

the mmW electronic frequency region and amplified to drive mmW MIMO antenna elements.

Connecting devices range from massive capacity premises, such as education campus data centers and direct access by academic departments, to machine-to-machine communications, fiber to the premises (FTTP), and fiber or optical to the antenna (FTTA/OTTA). The latter is shown in figure 2.8, in which the data communication link is delivering data to the antenna at which the optical signals are received and mixed with a local oscillator laser in the photodetector. The frequency of the LO is tuned to a frequency difference with respect to that of the data carrier of the millimeter wave (mmW) center frequency, e.g. 58.6 GHz. Thus there are opto-electronic conversions of optical signals to the mmW band. The mmW power amplifier delivers to the antenna element, a millimeter sized horn antenna embedded in a printed circuit board. This is the optical to MIMO antenna for a mmW band wireless carrier. Thus there would be a bank of receivers and mixers and mmW RF electronic drivers. A bank of photonic phase shifters is used to generate a pattern of phase delay so that beam steering can be implemented. These phase shifters must be modulators in integrated optic structure so that fast and high speed beam steering can be performed. Furthermore these OTTA are passive optical networks for distribution and hence share capacity and cost. The integrated photonic processor can be implemented in silicon photonics so that the cost can be sufficiently low. Beam steering phase modulators can be embedded using a p–n junction by ion implantation, operating in junction reverse mode in which speeds can be in tens of gigahertz.

With the 7.0 GHz bandwidth of the free-license band and 500 MHz per sub-band for data allocation, high level complex modulation can be used, as in the common

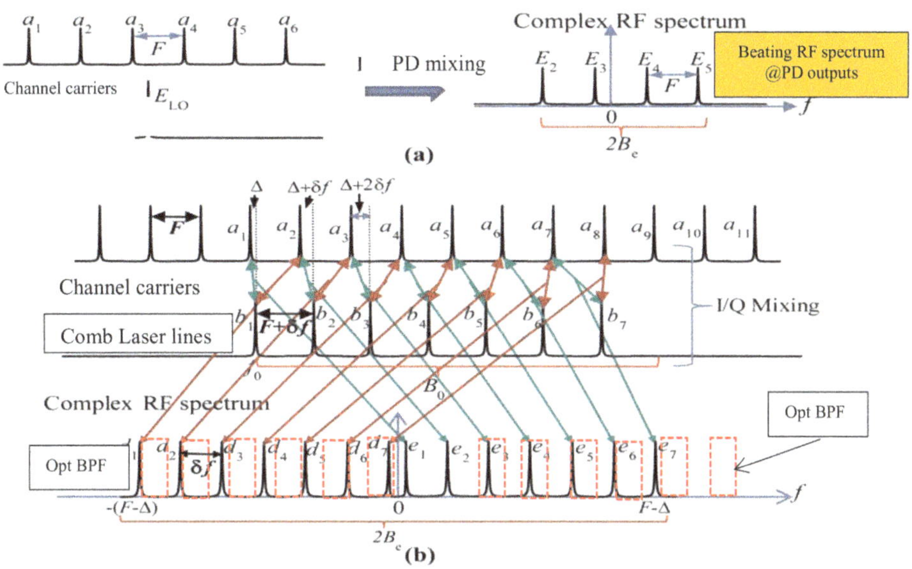

Figure 2.9. Comb sub-carriers and mixing in multi-super-channel transmission with mixing of multi-carrier signal channels and a comb source. Complex RF generated channels under an effective passband of optical filters. (a) Comb laser spectral lines as the source and local oscillator, and (b) beat frequency spectra in the electrical RF domain.

RF wireless communications. Thus more than 10 Gbps can be delivered to MIMO antenna elements. The frequency comb lines and the mixing spectral distribution in both optical and RF domains are shown in figure 2.9. Such a shift left or shift right of the spectra can be adjusted and designed based on the modulation of a single sideband or double sideband. With an antenna gain of about 22–0 dB for a 56 × 128 element array of the MIMO antenna, we can deliver more than 1.0 Tbps capacity, covering an area with a radius of 500 meters, by beam steering to clients of a crowded stadium.

Thus the current transport layers have been greatly migrating to lower latency and high capacity delivery to users via cloud broadband (cloud-BB) networks, from which direct access to wireless mobile devices can be implemented. Such dense-user environments include shopping centers, sport stadia, social gatherings, etc, with direct 3D video transmission/broadcasting at close up viewing.

2.2.2 Optical transport technology

2.2.2.1 Ring and tree networks

The transport of networks for 5G must evolve from the traditional ring networks to tree networks. The existing infrastructure in the current carrier networks deployed over the years by the carriers offers significant advantages to the flattening of the traditional telecoms carriers. That is, the ring configured structure can be modified to connect nodes to nodes, forming a multi-tree architecture with optimized latency and distributed MEC depending on capacity demands of 5G mobile users. In such a tree optical transport network the precision of the clock is much less than that

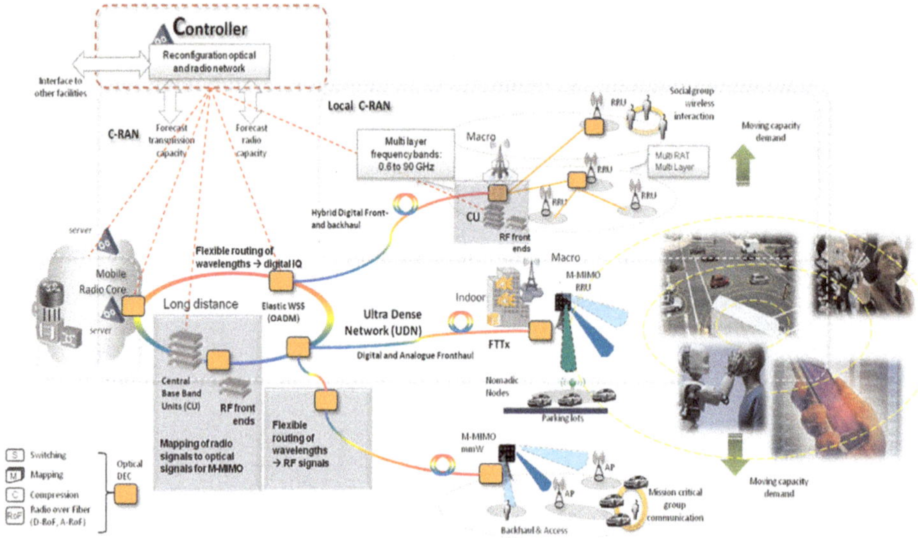

Figure 2.10. Generic view of 5G transport networks in a densely populated city, interconnected long-distance and user concentrated environments, as well as machine-to-machine IoET.

accumulating in ring network nodes to nodes by at least 4 to 5 times, hence the stability of tree networks. Tree networks minimize the latency to at least a few times less than that occurring in the ring node by node accumulation. The latency in tree networks can be reduced to less than 10 μs. For distributed RAN the latency can be less than 100 μs for tree networks. This latency is a few times longer for such a 3–4 node ring (figure 2.10).

We now touch on the modulation techniques and transmission systems required for low cost 5G access networking. The transmission systems for long-distance core and interconnection of data centers and core network nodes are covered in several papers published over the last decade (see for example many references given in [18]), in particular on DSP-based optical receiving systems in either coherent or direct detection (figure 2.11).

The standardization of 200 G/400 G[1] involves the PAM4 multi-level format with a baud rate of 28 GBd/56 GBd and multi-lanes to form such a high bit rate transmission line. The aggregate bit rate per λ (wavelength located for each accumulated channel) can be expected to be higher if other modulation formats can be applied.

2.2.2.2 Data transmission and transmission techniques

The discrete multi-tone (DMT) transmission technique using frequency-domain subcarriers modulated by higher level modulation can lead to reaching speeds of more than 200 Gbits s^{-1} (Gbps) per λ, thus for four lanes or eight lanes (optical carriers)

[1] LAN/MAN Standards Committee of the IEEE Computer Society *IEEE Standard for Ethernet: Amendment 10: Media Access Control Parameters, Physical Layers and Management Parameters for 200 Gb s^{-1} and 400 Gb s^{-1} Operation*, 6 December 2017.

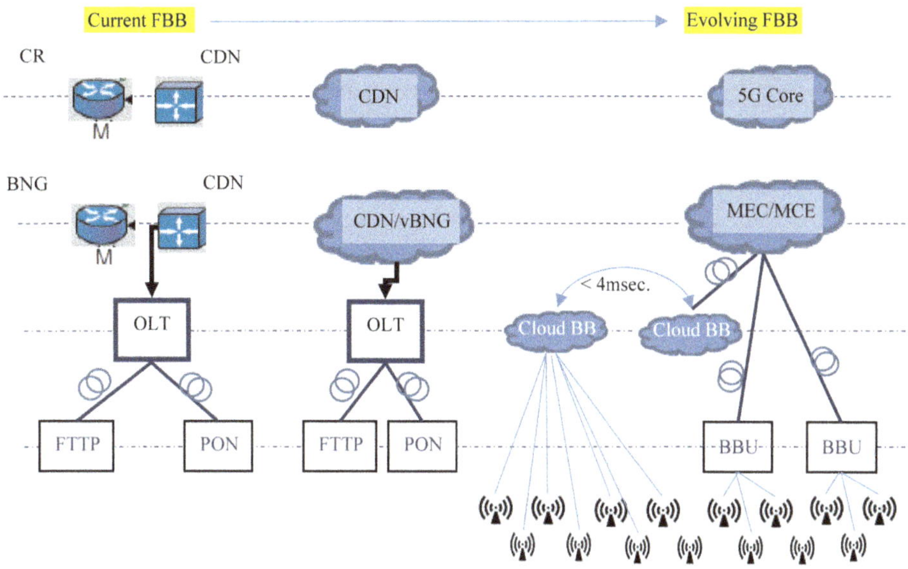

Figure 2.11. Evolution of current transport to cloudified networks for 5G. CDN = content delivery network; FBB = fixed broadband; CDN = current delivery network; vBNG = variable business network group; OLT = optical line terminal; MCE = mobile cloud engine; BBU = baseband unit; FTTP = fiber to the premise; PON = passive optical network.

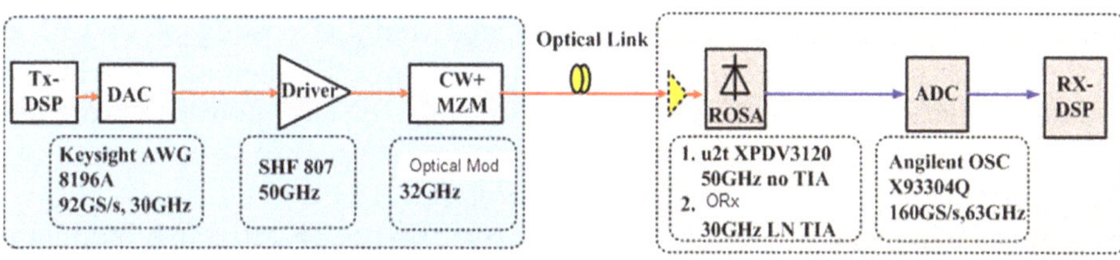

Figure 2.12. Optical transmission link for >200 Gbits s^{-1} per lamda (optical carrier) for 5G access capacity delivery.

one can obtain more than 1.0 terabits per transmission module over short distances for 5G access networking. Figure 2.12 depicts a transmission set-up for more than 200 Gbps per λ in the C-band. Orthogonal multi-sub-carriers are generated from an arbitrary waveform generator of bandwidth higher than 33 GHz and a sampling rate of 160 GSamples s^{-1}. The sub-carriers are modulated by PAM4 formats and then fed into a wideband driver to drive the optical modulator whose bandwidth is 32 GHz with an extension ratio at a biasing point of 7 dB in a completely linear region of the voltage–intensity transfer characteristics of the modulator. Figure 2.13 shows the bit error rate (BER) versus the transmission capacity of an optical carrier channel of an aggregate terabits s^{-1} link whose platform is shown in figure 2.12. DSP processing with forward error correction of 7% is able to recover the data channel without error.

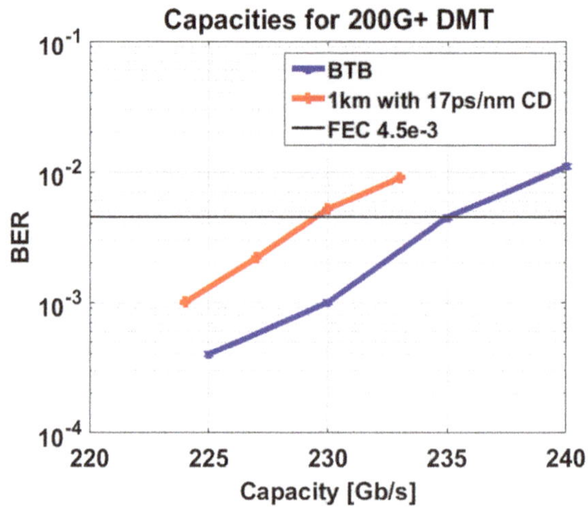

Figure 2.13. Bit error rate versus transmission capacity of a channel (per single wavelength λ) of a terabits s^{-1} transmission link.

Thus the most common and promising modulation formats are either the PAM4 optical domain with direct detection or QAM16 in a coherent DSP-based receiver. For much higher bit rates and direct detection the discrete multi-tone modulation in the frequency domain can offer higher than 200 Gbps with optical and electronic components of limited bandwidth around 34 GHz. Higher capacity can be reached with coherent transmission techniques provided that the costing of optical and electronic components are lower or much more novel coherent receiving and modulation techniques are devised, taking advantage of DSP power in processing.

2.2.2.3 Frequency-domain orthogonal modulation

Orthogonal frequency division multiplexing (OFDM) is an attractive modulation format that has recently received a lot of attention in the practical and R&D fiber-optic communities [19]. The main advantage of optical OFDM is that it can cope with a virtually unlimited amount of inter-symbol interference (ISI). In high speed optical transmission systems ISI is caused, for example, by chromatic dispersion (CD) and polarization mode dispersion (PMD), which are serious issues in long- and short-distance systems whose bit rate is higher than 56–100 GBd, in particular at these ultra-high rates and with multi-level modulation in the optical domain. In this sub-section some of the basics of an optical OFDM system are provided and high speed transmission experiments up to 122 Gbps are briefly presented (figure 2.14).

In principle, the basic concept behind OFDM is the division of a high bit rate data stream into several low bit rate orthogonal sub-carrier streams. In general, the sub-carriers are generated in the electronic digital domain, therefore these systems typically consist of many sub-carriers (typically more than 64 and should be in a factor of base-2 and exponent N). In these systems, channel estimation is realized by periodically inserting training symbols. In fiber-optic transmission systems, the OFDM systems where the sub-carriers are generated in the optical domain are also

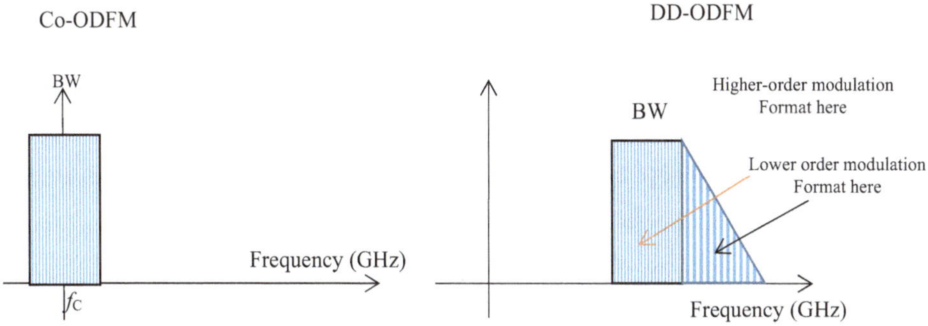

Figure 2.14. Spectra of optical OFDM—Co-OFDM and DD-OFDM. A sharp roll off at the spectrum boundary leading to compact channels multiplexing with a lower order of modulation format in the 3 dB to 6 dB band to compensate for lower sensitivity due to band limited components.

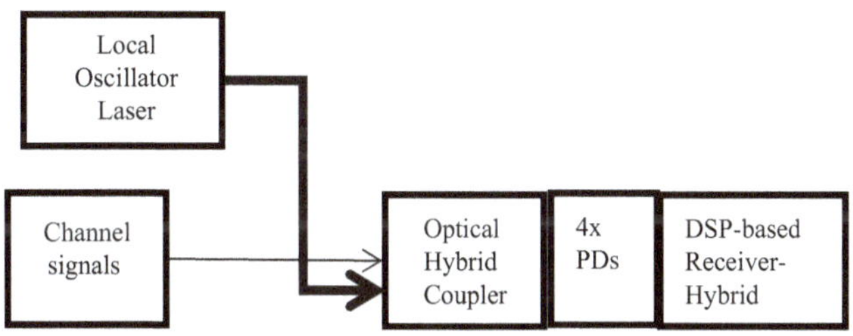

Figure 2.15. Schematic of an optical receiver for Co-OFDM.

created. These systems are sometimes referred to as coherent WDM systems [19]. Coherent WDM systems typically have few sub-carriers and do not use training symbols, but rely on blind channel estimation instead. In optical OFDM systems, the front end of the transmitter consists of an optical modulator, where the OFDM signal is unconverted to the frequency of an optical carrier. The front end at the receiver consists of either a coherent (Co) or a direct detection (DD) scheme. DD-OFDM is realized by sending the optical carrier along with the OFDM band so that direct detection with a single photodiode can be used at the receiver. In a Co-OFDM system, the optical carrier is suppressed at the transmitter and a local oscillator (LO) and an optical hybrid is required, as shown in figure 2.15. The superior performance of Co-OFDM with respect to the optical signal-to-noise ratio (OSNR) requirements, polarization mode dispersion (PMD) tolerance and spectral efficiency makes it an excellent candidate for long-distance transmission systems, whereas DD-OFDM, which requires fewer components at the receiver than Co-OFDM, is more suitable for cost-effective short reach applications. The spectra of both these two frequency multiplexing schemes are shown in figure 2.16. The DSP processing of DD-OFDM is much simpler as only the real part of the OFDM signals are required to be generated (figure 2.15). The alternative name for DD-OFDM is discrete multi-tone (DMT).

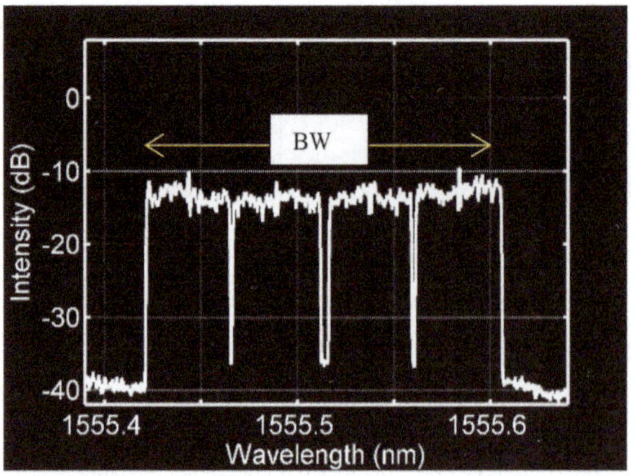

Figure 2.16. Optical spectrum of 122 Gbits s^{-1} OFDM signal, intensity versus the wavelength of four channels with a total bandwidth (BW) of about 28 GHz or $28 \times 0.8/100 = 0.224$ nm.

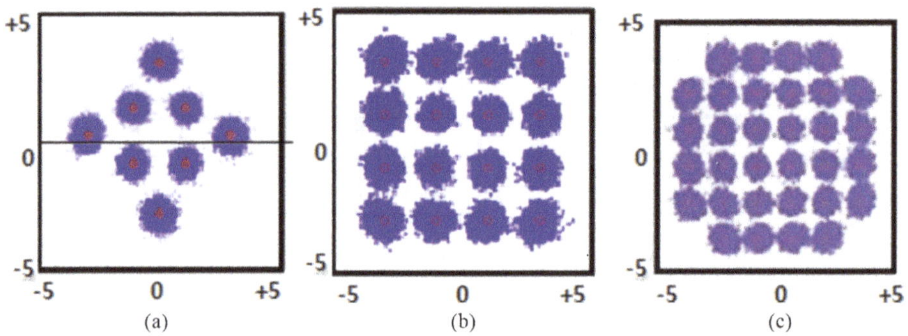

Figure 2.17. Signal constellations: (a) 8QAM, (b) 16QAM and (c) 32QAM.

A recent 100 Gbits s^{-1} transmission experiment has been reported recently. For both Co-OFDM [20] and DD-OFDM [21, 22] systems, 100 Gbps transmission experiments could be demonstrated. In order to increase the bit rate to the 100 Gbps level, polarization division multiplexing (PDM) is indispensable for a reduction of the signal bandwidth which determines the required bandwidth of components in the transmitter and receiver (figure 2.16). An ADC and then a DSP can be used in such a PDM-OFDM, and the polarization components can be separated digitally by multiple-input and multiple-output (MIMO) processing [23]. In order to apply the MIMO processing in DD-OFDM systems, an optical hybrid at the receiver [21] or a set of two carriers together with specially designed training signal patterns [22] could be employed. One group reported CO-OFDM transmission of 10 × 122 Gbps over 1000 km SSMF without any dispersion compensation [20]. In this experiment, the 122 Gbps OFDM signal consists of two polarization-multiplexed signals with four sub-carrier-multiplexed OFDM bands each. Together with a non-rectangular 8QAM constellation (figure 2.17(a)) used for symbol mapping, the 122 Gbps

OFDM signal can be packed in a 23 GHz bandwidth as shown in figure 2.16. At the receiver, the signal is coherently detected with an optical hybrid and an external cavity laser. A real-time digital storage oscilloscope is used to sample the outputs of the optical hybrid and the acquired data are post-processed off-line. Polarization reversed rotating can be realized through MIMO processing and an RF-aided phase noise compensation is also possible [24]. After 1000 km transmission, the obtained BER values for all channels are well below the threshold of a concatenated forward error coding (FEC) code with 7% overhead. Along the whole link, the OFDM signal is continuously detectable, demonstrating a dispersion tolerance of more than 18 500 ps nm^{-1}. Such a large dispersion tolerance is attractive for high speed transmission systems because it can eliminate the required inline dispersion compensator.

2.2.2.4 OTTA

The transport of data channels from cloud-RAN or D-RAN to massive parallel antennae, such as MIMO antenna elements in the mmW regions (58.6 GHz with 7 GHz bandwidth or in the 90 GHz band) shown in figure 2.8, are critical for increasing the wireless delivery capacity, but are yet to be exploited. One of the techniques we propose is the use of OTTA, with beating in the optical channels which is modulated in the baseband with an optical local oscillator whose frequency is different to the optical carrier channel by a millimeter wave frequency, say 58.6 GHz. The 7 GHz free-license band in this 56.8 GHz band is very attractive for delivering multi-Gbps (gigabits s^{-1}) over the 14 sub-bands, each with a bandwidth of 500 MHz and multi-level modulation. The number of elements of MIMO antennae can be 16 by 16 or 8 by 32. The antenna gain is sufficient to offer several Gbps per user over more than 200 meter wireless transmission.

2.2.3 Basic rates, capacity and server clusters for intra- and inter-DC connections

The transmission rates for interconnecting DCs must be as high as possible and the distance should be at least 40 km to 80 km of SSMF. The baud rate can reach 112 GBd so that for Co-OFDM the aggregate rate for two polarization channels of 16QAM would lead to 800 Gbps per wavelength with two polarized channels. Thus for direct detection PAM4 can be employed with a shorter transmission distance of 2 km to 10 kms for intra-DC networking.

Then the frequency-domain OFDM or DMT can offer much more compact channels in multiple mutiplexed channels with an increased number of sub-carriers. The band limited components can be used with their 6 dB band to increase the channel rate to >200 Gbps with only a 40 GHz 6 dB bandwidth of optical modulators and receivers.

We note that both the time and frequency-domain transmission systems require electronic digital signal processing. At 112 GBd the sampling rate of these processors is quite high and thus requires significant power consumption. These limitations and difficulties can be avoided by using photonic signal processors.

2.3 Exabits s^{-1} integrated photonic interconnection technology for flexible data-centric optical networks

Optical networking is evolving from classical service-provider based and data center based internetworking environments with massive capacity, hence requires novel optical switching and interconnecting technologies. We present proposed distributed and concentric data center based networks and the essential optical interconnection technologies, from the photonic kernels to electronic and opto-electronic server clusters, in both passive and active structures. Optical switching devices and integrated matrices are proposed, composed of tunable (bandwidth and center wavelength) optical filters and switches as well as resonant micro-ring modulators (μRM) that are switching and spectral demux/mux, for multi-wavelength flexible bandwidth optical channels of aggregate capacity reaching Ebps. The design principles and some experimental results are also reported.

2.3.1 Introductory remarks

Recent advances in information technology (IT) and the diffusion of broadband, fixed and radio connectivity [25, 26], and the significant increase of the transmission bit rates from 100 G to Tbps super-channels, are leading to an aggregate DWDM capacity approaching and possibly surpassing 100 Tbps per fiber. Hence the dramatic reduction of hardware costs per bit and the wider availability of open source software are creating access to innovative design and restructuring of the optical and radio connectivity, and metropolitan and long-distance transmission networks [27–31]. The emerging 5G wireless networks, their integration with cloud-based architectures [32] and demands on data services from smart mobile phone users will put tremendous pressure on the networking of data center to be structured in both distributed and concentric topologies.

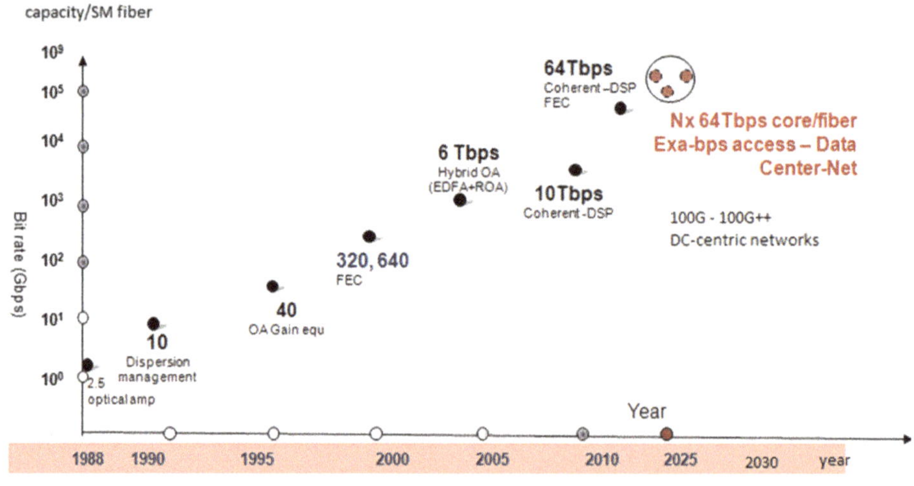

Figure 2.18. Evolution of transmission capacity in the past and predicted for the next 40 years.

The baud/symbol rate can now reach 56 GBd together with multi-level modulation formats and either coherent or direct reception, respectively. This enables the aggregate bit rates per single carrier to reach 400 Gbps, hence a capacity of several Tbps (see figure 2.18) if using multiple wavelengths with Nyquist pulse shaping or efficient spectrum packing in both long-distance and access networks.

Data servicing thus demands optical switching and routing for distributed and concentric topologies so as to overcome the limitations of electronic routers. Thus modern optical switching and routing systems must be able to offer: (i) non-blocking switching; (ii) flexible channel bandwidth filtering at the input and output ports with minimum cross-talk noise; (iii) fast switching speed; and, last but not least, (iv) the channel counts must be high, reaching a total capacity of about Ebps.

Recent advances in photonic devices permit optical switching system designers to consider this huge possibility in the implementation of compact Ebps optical switching systems, or systems-on-a-chip (SoCs). These devices include micro-ring resonators (MRR), X-switches, Mach–Zehnder interferometers (MZI) and cascaded multi-micro-rings (MMRs) [33, 34], etc. Micro-ring resonators (μRS) act as wavelength mux/de-multiplexers and switching/modulation simultaneously, in particular when the injection and depletion of impurity carriers are employed to manipulate the refractive index and hence phases of the lightwaves guided in Si photonic waveguides. The X-over switches are based on the principle of manipulating of the refractive index by carrier injections, hence altering the waveguide structures to divert the lightwaves in different directions. The MZI is necessary for cascading with ring resonators as these two devices complement each other in terms of offering a zero or depletion of guided lightwaves at a certain phase state that is different between the parallel paths of the interferometers. On the other hand, the ring resonators produce a maximum intensity when the total accumulated phase of the guided lightwaves reaches a multiple 2π, i.e. operating at resonance. This depleted interference and the resonance phenomena are equivalent, respectively, to the zero and pole of an electrical network or in the z-plane of the transfer function (optical transmittance) of a sampled optical network which is represented in terms of the delay sampled variable z [35].

Optical switching systems have been under extensive research over the years, employing guided wave structures such as $LiNbO_3$ [36–39]. Recent advances in Si photonics offer opportunities for the integration of both CMOS electronic analog and digital circuits, with photonic devices attracting significant interest in the SoC approaches to minimize the cost as well as increasing the input and out port counts. This SoC would fulfill the dream of integrated optics that was first proposed by S E Miller in 1969 [40]. The total capacity that a silica standard single mode fiber (SMF) can carry is about 100 Gbps, and even higher if a vestigial sideband is used and only in the C-band. So for Ebps we expect 10 000 fiber ports to be employed. It is not difficult to estimate this massive capacity requirement with 100 $\times$ 100 fiber connections at the two main ports of an optical matrix of north–east–west–south ports or an interconnection of smaller matrix size interconnection optical chips. It is this massive interconnection technology that we wish to overview in this section.

In this sub-section we present an overview of the following. (i) Data-centric networks comprising distributed data centers and data center concentric networks, and demands on optical networking whose capacity reaches Ebps under flexible multi-wavelength channels of optical transport networks, hence the general structures of active optical interconnecting technologies in both spatial and spectral domains with channels of aggregate bit rates ranging from 100 G to 400 G and to 10 or 32 Tbps. (ii) Integrated photonic technologies including MMR modulators/switches simultaneously performing as multiplexing and de-multiplexing functions, multi-X-over switches incorporating MMRs as the basis elements for multi-Tbps and Ebps interconnection matrices. The switched and transmission optical paths can handle the base baud rate variable from 100 G to 400 G and then Tbps, thus flexible bandwidth channels or flexible optical transport networks. The design of integrated tunable bandwidth optical filters for incorporating with X-switches is presented for forming Ebps optical interconnecting matrices. (iii) Finally, the fundamental designs of ultra-compact flexible filters and switching integrated components based on Si photonics for Ebps active interconnection are presented. Experimental results on the multi-channel transmission and performance of optical switching matrices and the effects on that of data channels are also reported.

2.3.2 Ebps optical network topologies

Figure 2.19 shows the evolutionary network structure strategy that will be emerging in the near future. This type of network exhibits a greatly simplified control and routing mechanism for servicing the demands of the end users, in particular in the new broadband Internet. It is expected that all traditional (classical) telecoms carrier networks will be flattened to a much more simplified structure in order to meet the challenges of such significant trends expected in the coming decades of the twenty-first century. It is expected that in terms of bandwidth consumption, the average bandwidth of the end user will be up to 1 Gbps (range: 100 Mbps–1 Gbps) with a

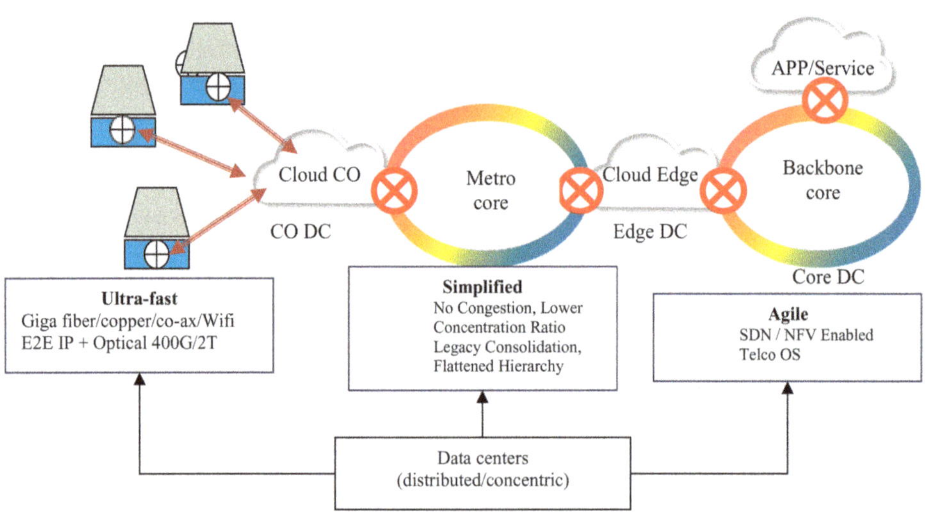

Figure 2.19. Evolutionary network modernization strategy.

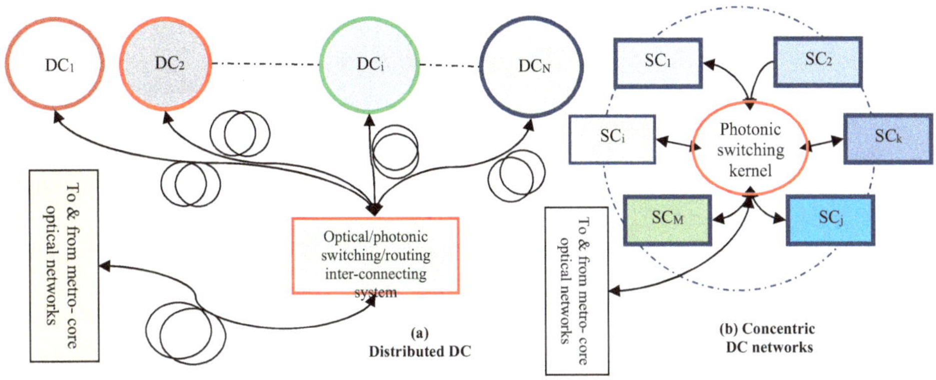

Figure 2.20. Generic structure of (a) distributed data centers interconnected via an optical switching center and (b) concentric data center consisting of server clusters interconnected via a photonic switching kernel.

coverage of 90% of households by high-bandwidth broadband service networking. This will lead to a demand for the transport of information to the multi-Tbps to be delivered to the edges of the networks in the metro core and access for the DCNs and classical flattened carrier networks.

Figure 2.20 shows the topologies of two typical data centers, the distributed and concentric types. The latter topology is very common and is expected to offer cloud operations for storage and servicing. For the former topology, the distributed interconnected data center networks (DIDCN), there are a number of medium range data centers placed closer to the mobile clouds or cloud radio access networks (CRAN) which are interconnected via an optical switching center. On the other hand, in a concentric data center (CDC) the transport to and from and the capacity of information to be switched or routed is very large and is reaching of the order of Ebps. The number of server clusters (SC) to be interconnected can reach an order of more than 10 000 with about 100 Tbps through each single fiber strand. Similarly, for distributed data center networking the capacity can reach this high with the expectation that there are thousands of data centers distributed in a region whose interconnecting distance satisfies the required dispersion tolerance and OSNR for a certain recoverable BER [41, 42].

The demands on the transmission links are different between the two topological networks. For DIDCN the transmission distance between the distributed center to the optical switching and interconnecting center would be in the distance from 2–10 kms, while this distance can be no longer than 300 m in a CDC. Furthermore, for both transmission link types, direct modulation and direct detection are used for economic reasons despite coherent detection and recovery of the reception of data channels transmitted from the core fiber networks, as shown in figures 2.20(a) and (b). It is thus expected that the data channels are recovered in the electronic domain and are then regenerated, switched and routed in the optical domain to appropriate server clusters or distributed data centers. This is due to the fact that routing at such an ultra-high baud rate (28 G or 56 G) would be fairly difficult and would have an unavoidably high probability of error in the electrical domain, due to cable/track frequency dependent losses.

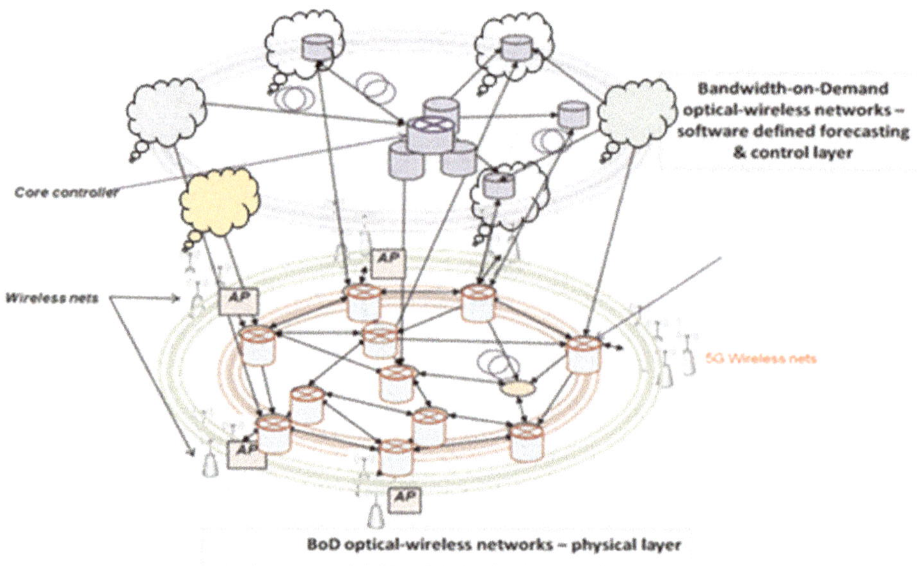

Figure 2.21. Converged flexible bandwidth optical transport structures and SDN control; a special case with 5G wireless access and optical core networks via several clouds.

Furthermore the current evolution of optical transport is focusing toward bandwidth on demand with flexible bandwidth of the overall super-channel, which can employ multiple modulated sub-carriers. The number of sub-carriers can vary and hence its effective bandwidth can be from 75 GHz for 400 G to 375 GHz for 1.0 Tbps, as shown in figure 2.21, with the network routing control right down to the optical physical layer, the optical software defined networking (OSDN) technique. In SDN, the control layer penetrates down to the physical layer. This control layer must also be matched and adapted to the topology and the capacity of the physical networks. Thus the photonic switching/routing systems should be simple even at ultra-high capacity operation.

2.3.3 Basic optical switches

Since 1969, integrated optics technology has witnessed intensive research on optical switching based on a number of basic phenomena, including electro-optic (EO) effects via an applied electric field thus the phase changes exerted on the lightwave paths, acousto-optic (AO) effects via elastic/perturbed stress/strain on crystal structure, the effective gratings for the Bragg or Raman–Nath diffraction regime for beam deflection/switching, and the magneto-optic (MO) effects to alter the optical phases. These are commonly used for moderate high speed optical switching. Optical switching can be implemented by employing a number of fundamental phenomena in optics, that include interferometers, cross-over waveguides by alteration of guided modes and resonance, and diffraction for deflection/spatial switching. We are interested in moderately high switching speed so the AO switches are not further presented here, only the EO or carrier injection (CJ) types are

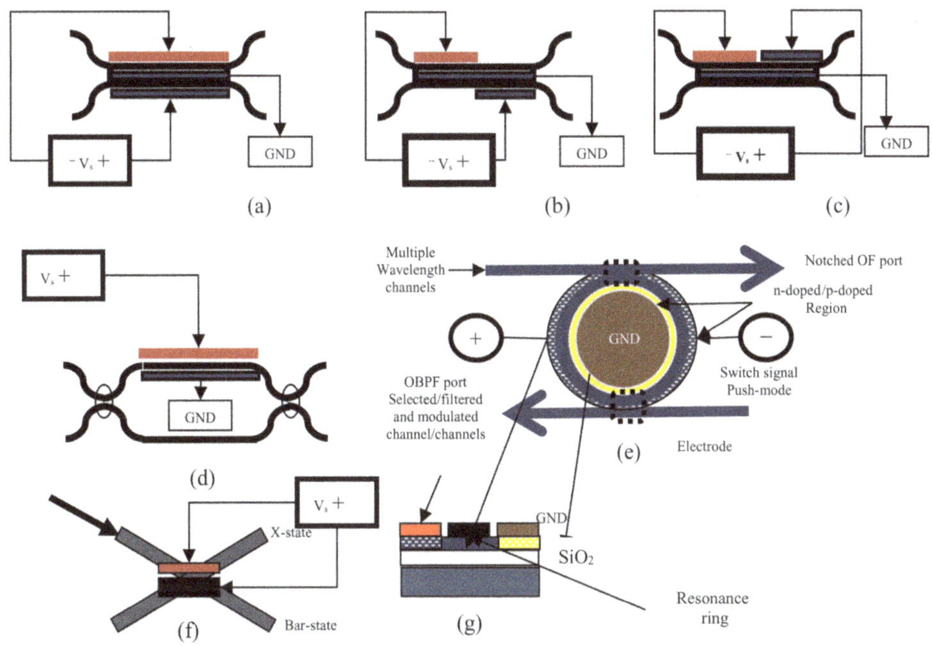

Figure 2.22. Integrated photonic switches employing carrier injection and change of refractive indices of waveguides and interferometers, X-over or resonators: (a)–(c) directional coupler switches in normal, differential and push–pull modes; (d) Mach–Zehnder interferometric (MZI); (e) RS; (f) X-over switch; and (g) cross-sectional view of Si waveguide and n-/p-doped region for carrier injection.

examined. The structures of these switches are shown in figure 2.22. It is noted that except for the resonant micro-ring of figure 2.22(e), all others are spatial switching with the passband being quite wide. Thus they cannot be employed in flexible grid optical switching or routing matrices, but must be integrated with other interferometric filters so that their passband can be tuned to the channel bandwidth as required. The passband of the micro-ring must also be wide enough to accommodate the aggregate bandwidth of the channels or super-channels. On the other hand, the switches shown in figures 2.22(a) and (b) are of directional coupler type whose cross coupling can be implemented by the carrier injection into one of the waveguides to either isolate (by increasing the refractive index) or spread the evanescent field to coupling the TE or TM field into the other waveguide branch. When the total phase reaches a quadrature state, then a complete switching occurs. For Si doped with electron impurity ΔN_e of $1.0e^{+17}$ cm^{-3}, the change is $-8.8 \times 10^{-22}\,\Delta N_e$ or $1e^{-3}$ [43] at an applied voltage of about 0.9 V and $1e^{-2}$ at 1.2 V, respectively. A similar change is also expected for the doping with holes as the minority carriers. The total change of the refractive index due to both electrons and corresponding holes is given as

$$\Delta n = \left[8.8 \times 10^{-22}\Delta N_e + 8.5 \times 10^{-18}\Delta N_h^{0.8}\right]. \tag{2.1}$$

Hence, a total length of about 487 and 244 μm, respectively, for a $\pi/2$ phase shift for type (a) and (b), that is $\Delta\beta_{\rm eff} \cdot L \cong k_0 \partial n_{\rm eff} \cdot L = \frac{\pi}{2}$, with β the propagation constant of the guided wave in the channel/ridge waveguide and L is the coupling length. Figure 2.22(f) shows the cross-section of the Si waveguide and the impurity doped regions as well as the electrodes to inject carriers into the guiding volume. For type (c) (figure 2.22(c)) the required phase for the cross bar state is π and via the 3 dB coupler at both ends requires the total length of the switch to be twice as long as type (a).

For the μRS the total resonance occurs when the total phase of the lightwaves around the ring equals a multiple of 2π, thus the free spectral range happens by an integer of this change. The ring can be set into resonance or out of this mode by carrier injection into the ring and a switching signal whose voltage level is sufficient to inject the impurities into the active ring waveguide area. Push–pull operation on the μRS can be implemented by differential driving signals, as shown in figure 2.22(e), to reduce the required driving voltage level or minimize the radius of the ring, as well as doubling the passband of the ring resonator.

2.3.3.1 *Tunable spectral bandwidth switch structures*

Photonic circuits, in particular optical filters, can be designed using techniques well known in digital signal processing with discrete variable z representing the delay. In photonics this delay can be represented as $z = e^{-j\beta L} = e^{-j\omega T}$ with β being the effective propagation constant of the guided wave in the optical waveguide, L is the propagation length, ω is the angular frequency of the lightwaves in a vacuum and T is the delay interval. Thus the photonic transfer function of an optical circuit can be written as a cascade of networks of transfer functions consisting of all poles and all zeroes as

$$\hat{H}(z) = \hat{A}\prod_{k=1}^{M}\frac{(z - \hat{z}_k)}{(z - \hat{p}_k)} = \hat{A}\frac{(z - \hat{z}_1)}{(z - \hat{p}_1)}\frac{(z - \hat{z}_2)}{(z - \hat{p}_2)}\dots\dots\frac{(z - \hat{z}_M)}{(z - \hat{p}_M)}, \tag{2.2}$$

where M is the total number of poles and zeroes of the optical networks, which is the same according to network theory. The poles and zeros are given by $\hat{z}_k$, $\hat{p}_k$. The values of the poles are corresponding to the resonant frequency at which the optical circuit resonates to obtain the greatest energy, such as those in a micro-ring when the total phase of the lightwaves circulating equals a multiple of 2π. Correspondingly, the zeros of an optical circuit occur when the output of the circuit becomes nullified, such as those in optical interferometers when the phases of the optical paths are in phase reversal with respect to one another. Thus we can observe that the filters can be constructed by cascading μRM and Mach–Zehnder delay interferometer (MZDI) incorporating phase shifters as shown in figure 2.23 [44, 45]. Theoretical design using the Butterworth filter in the discrete domain can be applicable to the photonic domain and the transfer function in parallel connection can be given as

$$H_{\mathrm{BW\_parallel}}(z^{-1}) = \frac{1.678719 \times 10^{-11}(1 + z^{-8}) - 6.714876 \times 10^{-11}z^{-2}(1 + z^{-4}) + 1.007231 \times 10^{-10}z^{-4}}{1 + 9453 \times 10^{-5}z^{-1} + 3.989407\, z^{-2} + 5.968278\, z^{-4} + 3.968334\, z^{-6} + 0.989463\, z^{-8}}. \tag{2.3}$$

Similarly as in parallel (interferometric), cascade branches of the interferometric form as

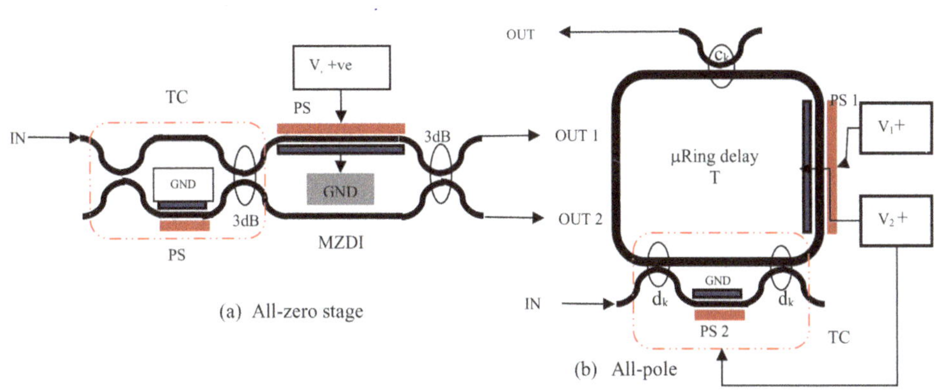

Figure 2.23. Schematic diagram of (a) all zero (interferometric) optical circuit and (b) all-pole (μRS) photonic stage. PS = phase shifter; TC = tunable coupler; MZDI = Mach–Zehnder delay interferometer. Bold lines represent optical waveguides.

$$H_{\text{BW\_cascade}}(z^{-1}) = 1.68 \times 10^{-11} + \frac{1 + 0.008\,z^{-1}}{1 + 0.004\,z^{-1} + z^{-2}} + \frac{1 - 0.008\,z^{-1}}{1 - 0.004\,z^{-1} + z^{-2}} + \frac{1 - 0.005\,z^{-1}}{1 + .002\,z^{-1} + z^{-2}} + \frac{1 + 0.005\,z^{-1}}{1 - 0.002\,z^{-1} + z^{-2}}. \tag{2.4}$$

The magnitude and phase responses as a function of wavelength can be obtained for an eigth-order filter, as shown in figures 2.23(a) and (b), respectively. Thus there are eight μR and three MZDI for this Butterworth filter. It is noted that a 3 dB passband of about 4 nm is observed for this design. A third-order filter can similarly be obtained with three μR without any difficulty [45] with a narrower passband. The bandwidth of these filters can be tuned by tuning the phase shifters and the coupling coefficients to accommodate for flexible channels (figure 2.24). Similarly, a multiple passband and sixth-order Chebyshev filter can be designed and implemented with the amplitude response shown in figures 2.25(a) and (b). Thus any photonic filter with a specific response can be designed and implemented using cascade MZDI and μR can be obtained with a tunable passband to accommodate flexible channels in Tbps.

The overall occupied area of these cascade optical circuits would be in the range of a few μm^2. Thus a dense optical switching matrix and flexible filters can be implemented in a reasonable area of a Si-based photonic integrated chip.

2.3.3.2 Photonic switching and routing kernels and systems

The speed of the tunable passband switch can operate at a reasonably high speed, which is contributed to by the rise-time taken for the injection or withdrawal of the impurity carriers from the guiding area. This is an electronic process and the frequency can reach a few tens of GHz. These filtering switches can be structured into a large matrix form, as shown in figure 2.26(a). The area limitation is due to the size of the pads of the electrode. In this schematic, the central part shows the

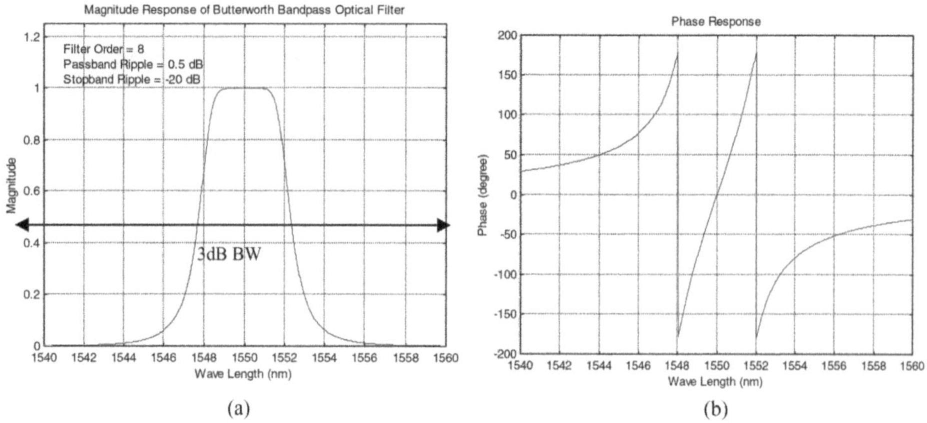

Figure 2.24. (a) Flat-top amplitude response of a linear-phase multi-band half-band photonic filter. Note the linear vertical scale. (b) Phase response of a Butterworth bandpass optical filter corresponding to (a).

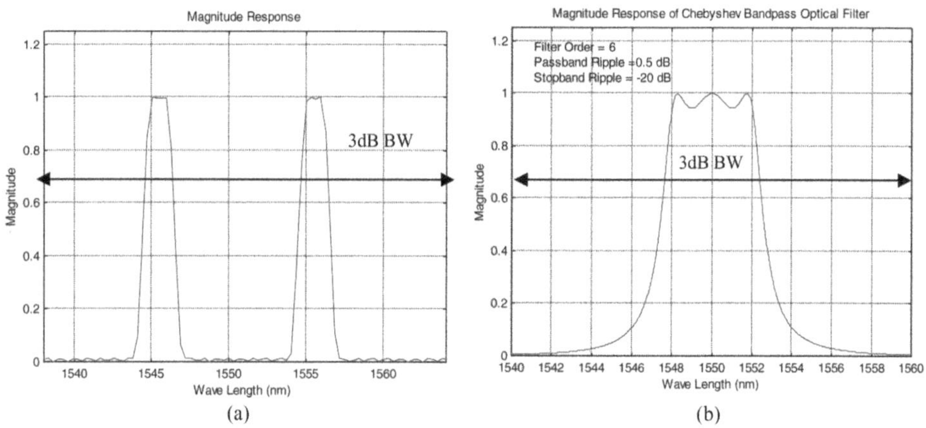

Figure 2.25. (a) Amplitude response of a linear-phase multi-band half-band photonic filter. Note the linear vertical scale. (b) Flat-top amplitude response of a linear-phase Chebyshev photonic filter with a ripple of 0.5 dB.

four-port switches whose bandwidth can be variable, as explained in section 2.3.3.1. The input ports and output ports must be coupled to multiplexers and de-multiplexers, which are most feasible by implementation in silica over Si technology that is also compatible with the CMOS process. Alternatively, the wavelength selective switches can be designed using μR, as shown in figure 2.26(b), in which three μRs are employed and each resonator is tuned to a specific wavelength and possibly tunable to a different bandwidth. The operational principles of this switching matrix are based on resonating to circulate the desired channel to a particular ring $\mu R_1/\mu R_2/\mu R_3$. The waveguide coupler WC_1 or WC_2 then couples the resonant channel to a specific output port. It is noted that the ring diameters are designed so that the WCs can be shared or separate depending on the switching control and designed matrix

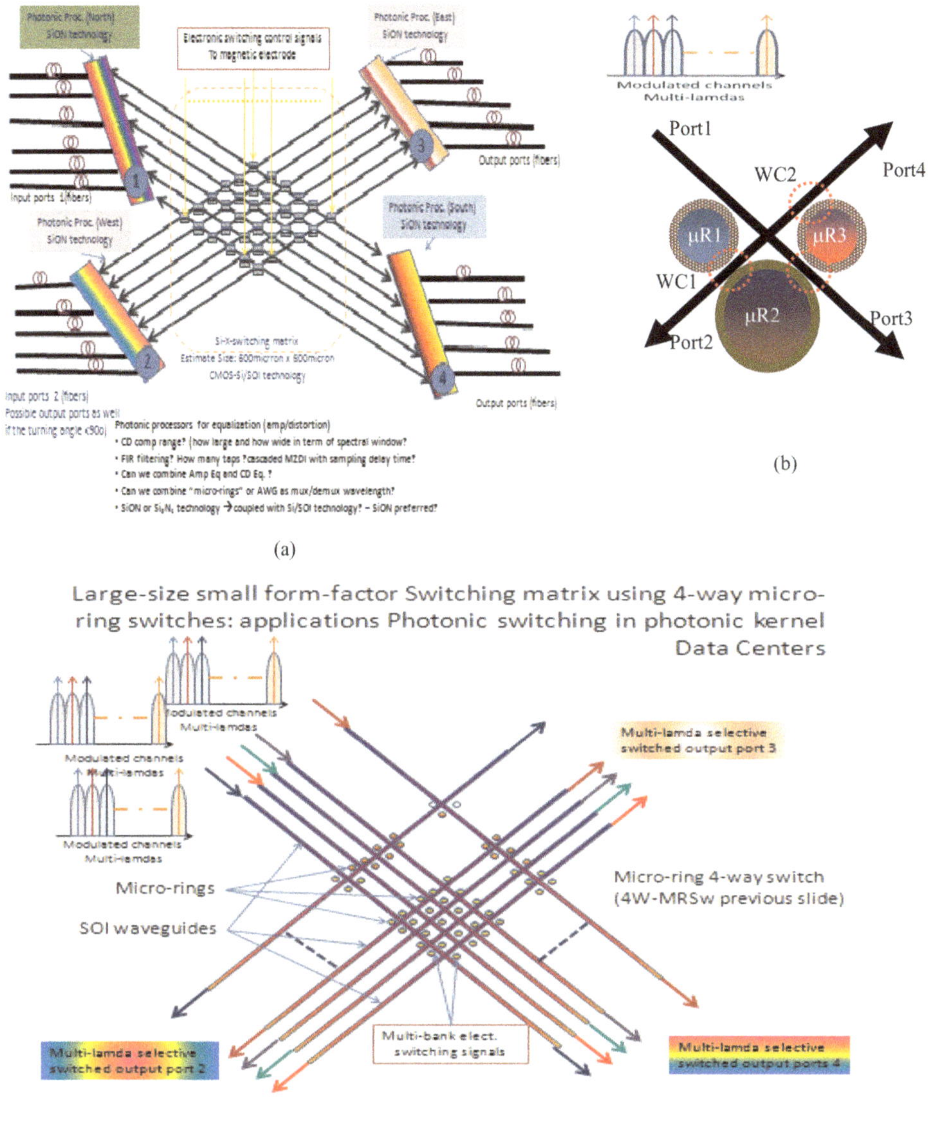

Figure 2.26. (a) A photonic processing–switching matrix using cascaded photonic circuit switches (Si/SOI MZDI and under field modulation/switching); (b) a four-way resonant selective optical router; and (c) a large form factor four-way switching matrix based on MRM routing (WC = waveguide coupler) incorporating photonic switching matrix employing μRS.

operations, as indicated by the circles of the WC in figure 2.26(b). However, the bandwidth of μR may not be that wide for super-channels and so these switching matrices formed by this type of X-μR of 2×2 type can only be used for massive interconnections whose channel bit rate may reach only 400 Gbps under multi-level modulation.

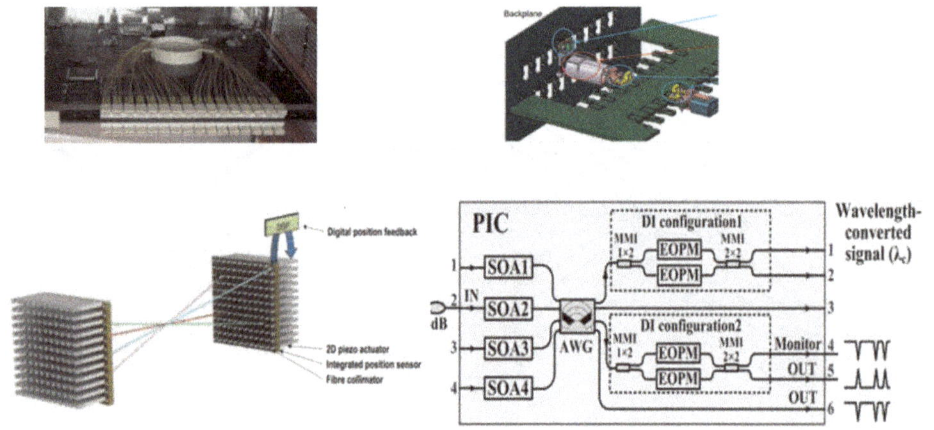

Figure 2.27. Schematic of a large scale optical switching matrix for both space and wavelength channel switching/routing.

2.3.4 Current technologies for optical switching and routing

An all optical free-space switching and routing matrix is shown in figure 2.27. The matrix size is of dimensions 384 × 384 spatial switching (Huber and Suhner, Switzerland) and 16 × 16 WDM switches in the C- and O-bands (TU Eindhoven) (WDM WSS switch ~6 dB). The spatial optical switching matrices are composed of a base unit with 12 ports in a single row, making an up to 384 × 384 optical switching matrix with only 1 dB loss per port-to-port switching (technical director: Nick Parsons, formerly of Polaris). The physical mechanism is by piezo-electric control of the mirrors deposited at the fiber ends, and rotating and alignment of the fiber to fiber ports. The free space is the medium through which the lightwaves are propagating from one fiber to the other. One of the disadvantages is that one channel fiber (of several colors) can be switched to one other output fiber, hence no multiple distribution. A schematic of an integrated optical switch matrix and routing of multiple wavelengths and spacing is shown in figure 2.28. The integrated switching/routing matrix can be of the Si photonic type and hence minimum power consumption in contrast to the free-space switching matrix shown in figure 2.27.

2.3.5 Remarks

This section has addressed the challenges and technological developments towards ultra-fast ultra-high capacity optical networks whose capacity may reach of the order of Ebps from the long-distance core to the metro core and access networks. Both core networks in long-distance and metro have employed DSP-based coherent reception and transmitters as well as transponders with bit rates of 100 G, 200 G, 400 G and super-channel Tbps in flexible grid spectra. Metro access networks would require low cost receiving systems, and hence direct detection technologies would be most appropriate. Given that DC centric networks are serving the huge demands of data intense consuming societies, traditional telecoms carriers are flattening their

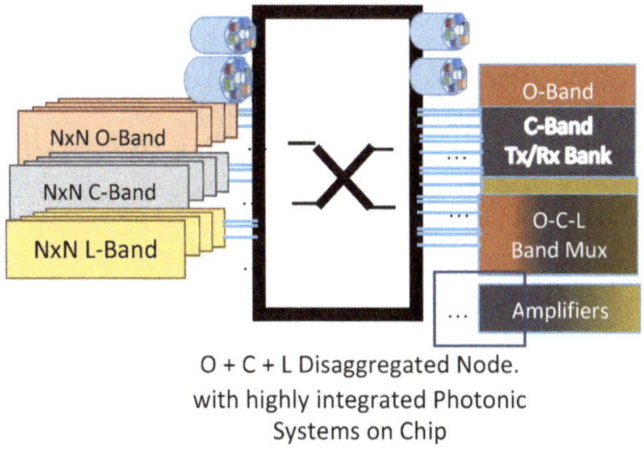

Figure 2.28. Schematic of an integrated switching/routing matrix.

networks so as to simplify and make them agile, as well as increasing their content services. The number of wavelength channels can now be planned to be about 92 with a frequency spacing of 50 GHz due to the improvement of optical amplifiers over the entire range of the C-band. Over this band the total transportable capacity can reach 100 Tbps per fiber. Several thousands of fibers are expected to be employed for the Ebps capacity and hence the optical switching and interconnection in such networks and intra- and inter-DC connections which are also proposed in sections 2.2 and 2.3. These optical routing systems would be deployed throughout the networks acting as flexible reconfigurable optical add/drop multiplexers (ROADMs) and optical cross connects (OXCs). The optical filters with sharp roll off characteristics given in section 2.3.3.1 allow the optical pulse shaping, e.g. Nyquist/raise cosine pulse shaping, that will allow high density packing of super-channels for such Ebps networks.

Integrated Si photonics are expected to offer the integration of optical components, both in the active and passive modes, for high speed transmitters and receivers under coherent or non-coherent reception and OXC as well as ROADM. The designs for such OXC/ROADM are given in section 2.3.4. Furthermore, integration of laser sources including comb laser structures on Si substrates will also be critical to obtain compact and low cost multi-Tbps systems for Ebps and even Zbps networks.

2.4 Concluding remarks

This chapter gives a brief description of the structures of DCN for both intra- and inter-DCs. The transmission techniques are also given so that the total capacity rate can reach higher than 800 Gbps and even higher than 1.0 Tbps. Thus in chapters 3, 4 and 5, both the DSP-based and PSP-based transmission systems are described in detail. A photonic processor for 5G mmW MIMO antenna beam steering in cloudified networks is provided.

References

[1] Bilal K *et al* 2013 Quantitative comparisons of the state of the art data center architectures *Concurr. Comput.: Pract. Experience* **25** 1771–83

[2] Noormohammadpour M and Raghavendra C S 2018 Datacenter traffic control: understanding techniques and trade-offs *IEEE Commun. Surv. Tut.* **20** 1492–525

[3] Al-Fares M, Loukissas A and Vahdat A 2008 A scalable, commodity data center 2 network architecture *ACM SIGCOMM 2008 Conf. on Data 3 Communication (Seattle, WA)* pp 63–74

[4] Guo C, Wu H, Tan K, Shi L, Zhang Y and Lu S 2008 DCell: a scalable and fault tolerant network structure for data centers *ACM SIGCOMM Comput. Commun. Rev.* **38** 75–86

[5] Bilal K, Khan S U and Zomaya A Y Green data center networks: challenges and opportunities *11th IEEE Int. Conf. on Frontiers of Information Technology (FIT) (Islamabad, Pakistan, December 2013)* pp 229–34

[6] Cisco 2010 *Cisco Data Center Infrastructure 2.5 Design Guide* (Indianapolis, IN: Cisco)

[7] Bilal *et al* 2014 A taxonomy and survey on green data center networks *Future Gener. Comput. Syst.* **36** 189–208

[8] Bilal K, Manzano M, Khan S U, Calle E, Li K and Zomaya A Y 2013 On the characterization of the structural robustness of data center networks *IEEE Trans. Cloud Comput.* **1** 64–77

[9] Guo C *et al* 2009 BCube: a high performance, server-centric network architecture for modular data centers *ACM SIGCOMM Comput. Commun. Rev.* **39** 63–74

[10] Costa P *et al* 2010 'CamCube: a key-based data center' *Technical Report* MSR TR-2010-74, Microsoft Research

[11] Li D *et al* 2009 FiConn: using backup port for server interconnection in data centers *IEEE INFOCOM 2009*

[12] Singla A *et al* 2012 Jellyfish: networking data centers randomly *9th USENIX Symp. on Networked Systems Design and Implementation (NSDI)*

[13] Gyarmati L and Trinh T A 2010 Scafida: a scale-free network inspired data center architecture *ACM SIGCOMM Comput. Commun. Rev.* **40** 4–12

[14] Manzano M, Bilal K, Calle E and Khan S U 2013 On the connectivity of data center networks *IEEE Commun. Lett.* **17** 2172–5

[15] Bilal K, Khan S U, Madani S A, Hayat K, Khan M I, Min-Allah N, Kolodziej J, Wang L, Zeadally S and Chen D 2013 A survey on green communications using adaptive link rate *Clust. Comput.* **16** 575–89

[16] Heller B *et al* 2010 ElasticTree: saving energy in data center networks *Proc. of the 7th USENIX Symp. on Networked Systems Design and Implementation, NSDI 2010 (April 28–30, 2010, San Jose, CA)*

[17] Matthias S, Gomes N, Juarer. A, Lange M, Leiba Y, Mano H, Murata H and Stoehr A 2018 Public field trial in a shopping mall of multi-RAT (60 GHz 5G/LTE/WiFi) mobile networks *IEEE Wirel. Commun. Mag.* **25** 38–46

[18] Binh L L 2017 *Optics and Photonics* (Boca Raton, FL: CRC Press)

[19] Shieh W and Djordjevic I 2009 *OFDM for Optical Communications* 1st edn (New York: Academic)

[20] Jansen S *et al* 2009 121.9-Gb/s PDM-OFDM transmission with 2-b/s/Hz spectral efficiency over 1000 km of SSMF *IEEE J. Lightwave Technol.* **27** 177–88

[21] Schmidt B J C, Zan Z, Du L B and Lowery A J 2009 100 Gbit/s transmission using single-band direct-detection optical OFDM *Optical Fiber Communication Conf. and National Fiber Optic Engineers Conf., OSA Technical Digest (CD)* paper PDPC3

[22] Qian D, Cvijetic N, Hu J and Wang T 2010 108 Gb/s OFDMA-PON with polarization multiplexing and direct-detection *Optical Fiber Communication Conf. and National Fiber Optic Engineers Conf., OSA Technical Digest (CD)* paper

[23] Jansen S *et al* 2008 Long-haul transmission of 16 × 52.5 Gbits/s polarization-division-multiplexed OFDM enabled by MIMO processing *J. Opt. Network.* **7** 173–82

[24] Jansen S *et al* 2008 Coherent optical 25.8-Gb/s OFDM transmission over 4160-km SSMF *J. Lightwave Technol.* **26** 6–15

[25] 2014 5G experimental facilities in EuropeNetworld 2020 ETP White paper

[26] Puttmtnam B J *et al* 2015 2.15 Pb/s transmission using 22 core homogeneous single mode multi-core fiber and wideband optical comb *ECOC2015 (Valencia, Spain)* PDP3.2

[27] Binh L N, Mao B N, Xie C S, Stonajovic N and Yang N 2015 Synchronous modulator incorporated re-circulating comb laser sources for superchannel optical transmission *Opt. Commun.* at press

[28] Zhou X *et al* 64Tb/s (640 × 107-Gb/s) PDM-36QAM transmission over 320 km using both pre- and post-transmission digital equalization *Proc. OFC 2010 (San Diego, CA, March 2010)* paper PDPB9

[29] Karinou F, Stojanovic N, Xie C S, Ortsiefer M, Daly A, Hohenleitner R, Kögel B and Neumeyr C 2015 28 Gb/s NRZ-OOK using 1530-nm VCSEL, direct detection and MLSE receiver for optical interconnects *Proc OFC 2015 (Los Angeles, CA)*

[30] Lee J, Kaneda N, Pfau T, Konczykowska A, Jorge F, Dupuy J-Y and Chen Y-K 2015 Serial 103.125-Gb/s transmission over 1 km SSMF for low-cost, short-reach optical interconnects *Proc. OFC2015 (Los Angeles, CA)*

[31] Rahman T, Rafique D, Spinnler B, Pincemin E, Bouëtté C L, Jauffrit J and Calabro S Record field demonstration of C-band multi-terabit 16 QAM, 32 QAM and 64 QAM over 762 km of SSMF *Opto-Electronics and Communications Conf. (OECC), (Shang Hai, China, June 28-July 2, 2015)*

[32] Liu J, Zhao T, Zhou S, Cheng Y and Niu Z 2014 Concert: a cloud-based architecture for next generation cellular systems *IEEE Wireless Commun.* **21** 14–22

[33] Ngo N Q and Binh L N 2006 Synthesis of tunable optical waveguides filters using digital signal processing techniques *IEEE J. Lightwave Technol.* **24** 3520–30

[34] Binh L N 2009 Tunable photonic filters: a digital signal processing design approach *Appl. Opt.* **48** 2799–810

[35] Binh L N 2007 *Photonic Signal Processing* (Boca Raton, FL: CRC Press)

[36] Deng K-L, Runser R J, Toliver P, Glesk I and Prucnal P R 2000 A highly scalable, rapidly-reconfigurable, multicasting-capable, 100-Gbit/s photonic switched interconnect based upon OTDM technology *IEEE J. Lightwave Technol.* **18** 1892

[37] Prucnal P R, Baby V, Rand D R, Wang B C, Xu L and Glesk I 2002 All optical processing in switching networks *LEOS* http://photonicssociety.org/newsletters/oct02/prucnal.html

[38] Prucnal P R, Blumenthal D J and Perrier P R 2006 *Self-optical Switching with Optical Processing* (Berlin: Springer)

[39] Prucnal P R 1987 Integrated optics and optical switching *Proc. ECOC/LAN'87*

[40] Miller S E 1969 Integrated optics: an introduction *Bell Syst. Tech. J.* **48** 2059–69

[41] Karinou F, Prodaniuc C, Stojanovic N, Ortsiefer M, Daly A, Hohenleitner R, Kögel B and Neumeyr C 2015 Experimental performance evaluation of equalization techniques for 56 Gb/s PAM-4 VCSEL-based optical interconnects *Proc. ECOC 2015 (Valencia, Spain)*

[42] Karinou F, Prodaniuc C, Stojanovic N, Ortsiefer M, Daly A, Hohenleitner R, Kögel B and Neumeyr C 2015 Directly PAM-4 modulated 1530-nm VCSEL enabling 56 Gb/s/λ data-center interconnects *IEEE Photonics Technol. Lett.* **27** 1872–5

[43] Soref R A and Bennett B R 1987 Electro-optical effects in silicon *IEEE J. Quantum Electron.* **23** 123–9

[44] Ngo N Q and Binh L N 2006 Synthesis of tunable optical waveguides filters using digital signal processing technique *IEEE J. Lightwave Technol.* **24** 3520–31

[45] Poon A P, Luo X, Xu F and Chen H 2009 Cascaded microresonator-based matrix for silicon on-chip optical interconnection *Proc. IEEE* **7** 1216–37

Le Nguyen Binh

Chapter 3

Access and data center networking transmission technologies

3.1 Introductory remarks

This chapter focuses on the description of the functional principles and operation of the modulation formats and receiving techniques of incoherent and coherent reception so as to deliver the highest speed with limited transmission distance (figure 3.1). A schematic of the optical pre-processing of the coherent reception of polarization multiplexed optical channels is shown in figure 3.2, all of which can be implemented in integrated photonic structures, in particular Si photonic technology [1, 2].

Under coherent reception the aggregate bit rate can reach 800 Gbps per optical carrier with both polarizations of guided waves in association with the digital signal processing in the electronic digital domain, or effectively the baud rate is at 112 GBd. Coherent orthogonal frequency division multiplexing (OFDM) can also be employed to reach this high rate and beyond. Such 5G optical networking in distributed topology is shown in figure 3.1.

Then, for low cost and a high rate under the limited bandwidth of optical, opto-electronic and electrical components, the use of discrete multi-tone (DMT) or direct detection (DD-)OFDM, with a sufficient number of sub-carriers and the extended bandwidth of 0–6 dB and lower order modulation formats, can extend the bit rate per optical carrier to higher than 200 Gbps with additional capacity employed in the 3–6 dB bandwidth region and a lower order of modulation. Furthermore, we also describe the PAM4 technique in such transmission to 112 Gbps or 56 GBd per optical carrier with use of a minimum DSP sampling speed.

The aggregation of four to eight optical carriers leads to a total bit rate in Tbps that allows the ultra-high capacity link in the data center (DC) networking and access layer of the evolved 5G optical Internet. The typical distance is 2–10 km and the spectral window can be in the O-band (1310 nm window) or 1550 nm spectral window, or even the 850 nm window by a vertical-cavity surface-emitting

doi:10.1088/978-0-7503-2292-8ch3

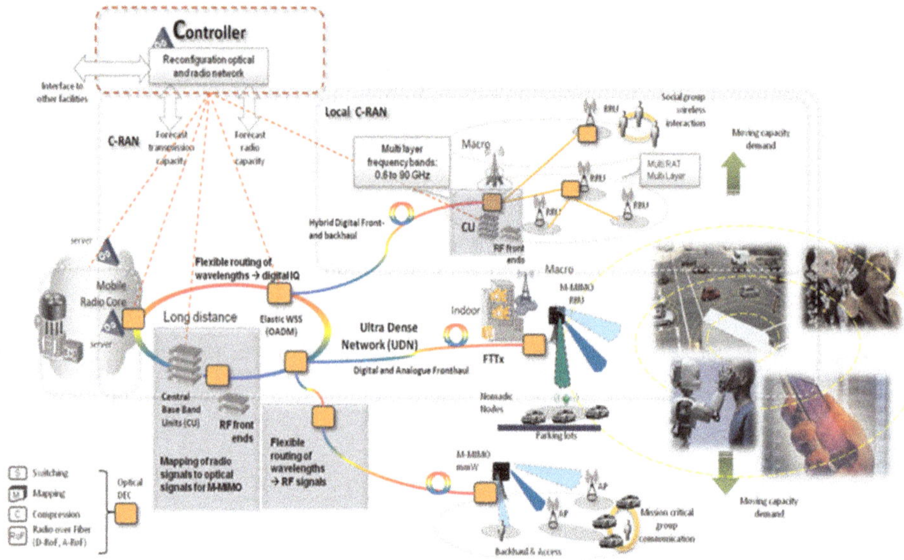

Figure 3.1. DC networking in the evolution of optical networks from core and metropolitan to 5G distributed networks.

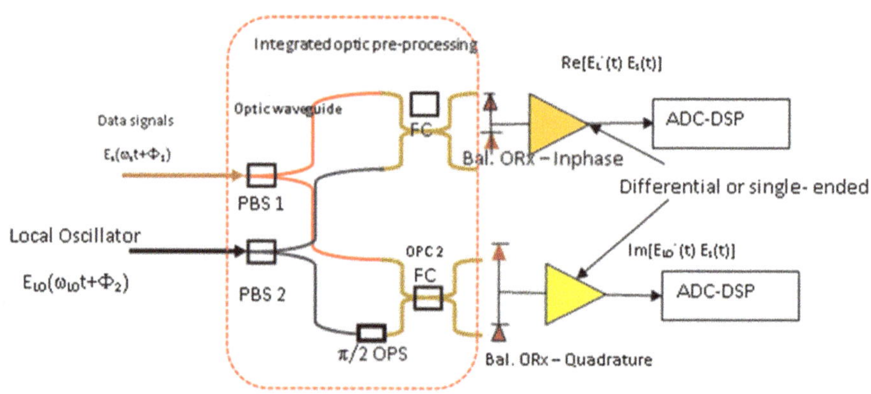

Figure 3.2. Schematic of a coherent receiver using a balanced detector and an integrated optic pre-processor for the *I–Q* components' phase and amplitude recovery, incorporating DSP and ADC. Im = imaginary part or quadrature phase component; Re = real part or in-phase component; PBS = integrated optic polarization beam splitter; ADC = analog too digital converter; DSP = digital signal processor; TIA = trans-impedance amplifier; ω_s, ω_{LO} = signal carrier and local oscillator frequencies; OPS = optic phase shifter; OPC = optical polarization combiner (same pol); TE = transverse electric (horizontal); TM = transverse magnetic (vertical).

laser (VCSEL). A typical commercial SFP module for 1.25 Gbps is shown in figure 3.3 at a cost per bit of about 10\$/1.25 Gbps or ~10\$/billion bits. This cost is expected to increase by ten times but for 100 Gbps in the near future. It is noted that the working spectral region is in the 1310 nm window using VCSEL, at which the dispersion effect is almost nil. Table 3.1 shows some current typical interfaces

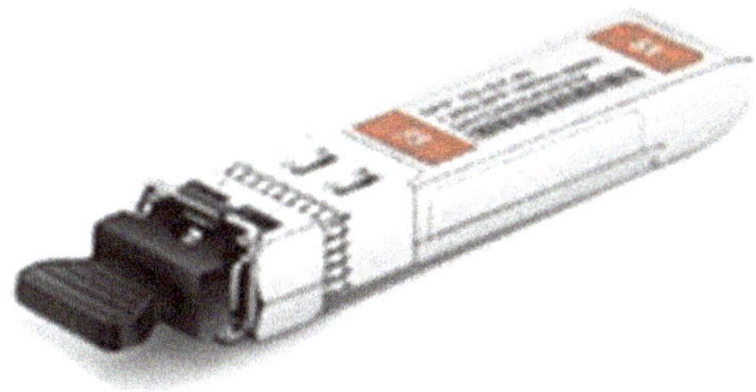

Figure 3.3. A commercial Gbps SFP size compatible device operating at 1310 nm with a 2 km transceiver module. Operating data rate up to 1.25 Gbps. Technology: 850 nm VCSEL laser transmitter; fiber: 550 m with 50/125 μm MMF, 300 m with 62.5/125 μm MMF; operating conditions: single 3.3 V power supply and LVTTL logic interface; pluggable SFP footprint duplex LC connector interface; class 1 FDA; commercial temperature range: 0 ~ +70°C; compliant with SFP MSA specification; built-in digital diagnostic functions, including optical power monitoring (2019 price ~$10). See also table 3.1 for more types and standards.

Table 3.1. Typical compact transmission interfaces with high capacity up to 5 Gbps. 100 G modules are expected to be the same size with integrated optic devices.

Type	Wavelength	Transmission distance	Fiber waveguide link	IEEE 802.3 Standard, remarks
1000BASE-SX	850 nm	500 m	50 μm/125 μm fiber pair MMF	Clause 38
		220 m	62.5 μm/125 μm fiber pair MMF	
1000BASE-LX	1310 nm	5 km	9 μm/125 μm fiber	Clause 38
1000BASE-LX10 1000BASE-LH		10 km	pair SSMF	Clause 59, '1000BASE-LX'
1000BASE-EX	1310 nm	40 km	9 μm/125 μm SSMF fiber pair	no IEEE 802.3 Standard, negligible dispersion
1000BASE-ZX	1550 nm	80 km	9 μm/125 μm SMF fiber pair	no IEEE 802.3 Standard, negligible dispersion
1000BASE-EZX	1550 nm	120 km	9 μm/125 μm SMF fiber pair	no IEEE 802.3 Standard
1000BASE-BX10	1490 nm(-D)	10 km	9 μm/125 μm	Clause 59, CWDM with 1490 nm
1000BASE-BX	1310 nm(-U)		SMF fiber pair	'downstream' and 1310 nm 'upstream'
1000BASE-T	electrical	100 m	Twisted pair Cat-5e	Clause 40, SFP module

for non-coherent transmission links for DC networking and short haul low cost access networking. Note that any DSP would be in a separate electronic module in association with these modules. The principles of coherent transmission systems can be found in [3] and are given in section 3.16. Section 3.17 illustrates the operational principles of balanced reception.

3.2 DSP-based coherent optical transmission systems

A flow chart of the digital signal processing in a coherent DSP-based transmission system is shown in figure 3.4. This diagram is now introduced here to include an optical processing section and the sampled signals are then digitally processed in a DSP, including the clock/timing recovered signals fed back into the sampling unit of the analog to digital converter (ADC) so as to obtain the best correct timing for sampling the incoming data sequence for processing in the DSP. Any errors made at this stage of timing lead to high deviation of the bit error rate (BER) in the symbol decoder, shown in figure 3.4. Dual polarized channels are propagating in the same optical guided beam and arrive at the coherent receiver consisting of an optical pre-processor in which a local oscillator (LO) laser is mixing with the signals in a PD, a square-law detector. The beating signals resulting from the square of the summation

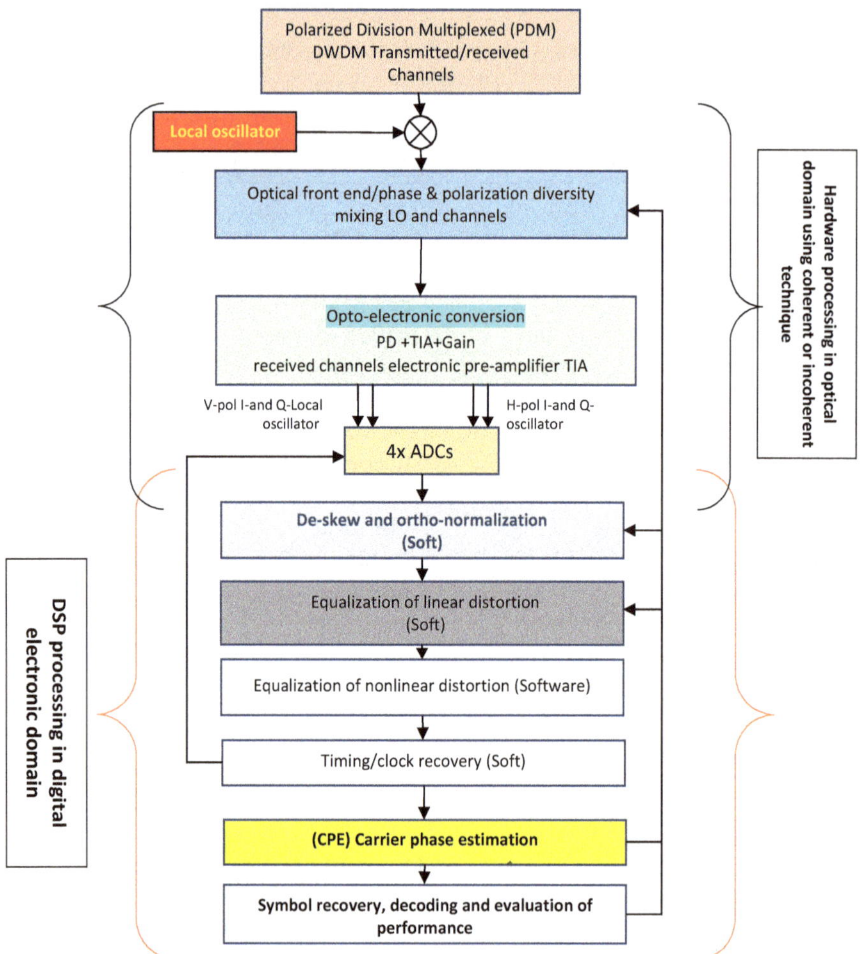

Figure 3.4. Flow of functionalities of DSP processing in a coherent optical receiver of a coherent transmission system with the feedback path diverted to ADC.

of the LO and the small signal level of the arrived channels is composed of three parts: (i) a dc term due to the intensity of the LO as the dominated term; (ii) an RF term as a multiplication of the amplitude of the LO and the signal with the frequency term as the difference between the carried signals and the LO frequency; and (iii) the frequency term as the amplitude of the summation of the frequencies of the LO and the signal carrier. The frequency band of the third term is too far away from the sensitivity of the PD and is thus eliminated. The dc term is also filtered by the low frequency roll-off of the opto-electronic (O/E) components, hence is negligible.

It is also noted that the vertically polarized channel (V-pol) and horizontally polarized (H-pol) channels are detected in the integrated optic pre-processing (IOPP) section, a hybrid coupler (HC) and a balanced PD pair (PDP). Their in-phase (I-) and quadrature (Q-) components are produced in the electrical domain with a signal voltage conditioned for the conversion to the digital domain by the ADC.

When several channels are transmitting in association with multi-carrier optical sources and compact bandwidth shaping, they are called super-channels. This technique is also important for carrying massive capacity in DC networking. The processing of a sampled sequence from the received optical data and photo-detected electronic signals passing through the ADC relies on the timing recovery from the sampled events of the sequence. The flowing stages of the blocks given in figure 3.4 may be changed or altered accordingly, depending on the modulation formats and pulse shaping, for example the Nyquist pulse shaping in Nyquist superchannel transmission systems.

This section describes the performance and impact of processing algorithms in optical transmission systems, short distance or long distance, employing coherent reception techniques over highly dispersive optical transmission lines, in particular the multi-span or single-span optically amplified non-DCM long distance. First, the QPSK homodyne scheme is examined, and then the 16QAM incorporating both polarized division multiplexing (PDM) in the optical domain, hence the terms PDM-QPSK or PDM-16QAM. We then expand the study to superchannel transmission systems in which several sub-channels are closely spaced in the spectral region so as to increase the spectral efficiency so that the total effective bit rate must reach at least 1.0 Tbps. Due to the overlapping of adjacent channels, there is the possibility that modifications of the processing algorithms are to be made. Further, the nonlinearity impairments on transmitted sub-channels would degrade the system performance and the application of DSP-based processing algorithms, such as back propagation algorithms, must be combined with linear and nonlinear equalization schemes so as to effectively combat the degrading.

A schematic of a balanced optical receiver is shown in the appendix, section 3.17.

3.3 Quadrature amplitude modulation (QAM)

Homodyne coherent reception requires a perfect match of the frequency of the signal carrier and the LO. The phase difference contributes significantly to the degradation of QAM systems as the phase orthogonality is the main feature of these modulation schemes. Any frequency difference will lead to phase noise of the detected signals.

This is the largest hurdle for the first optical coherent system initiated in the mid-1980s. In DSP-based coherent reception systems the recovery of the carrier phases and hence the frequency is critical to achieve the most sensitive reception with maximum performance in the BER, or evaluation of the probability of error. This section illustrates the recovery of the carrier phase for QPSK and 16QAM optical transmission systems. The constellations of these quadrature amplitude modulation schemes can be represented by a set of geometrical circles with different amplitudes. How would the DSP algorithms perform under the physical impairment effects on the recovery of the phase of the carrier?

Currently, several equipment manufacturers are striving to provide commercial advanced optical transmission systems at 100 Gbps employing coherent detection techniques for long distance backbone networks and metro networks as well. Since the 1980s, it has been well known that single-mode optical fibers can support such transmission due to the preservation of the guided modes and their polarized modes of the weakly guiding linearly polarized (LP) electromagnetic waves [4]. Naturally both transmitters and receivers must satisfy the coherency conditions of narrow linewidth sources and coherent mixing with a local oscillator, an external cavity laser, to recover both the phase and amplitude of the detected lightwaves. Both polarized modes of the LP modes can be stable over long distances so far, in order to provide the polarization division multiplexed (PDM) channels, even with polarization mode dispersion (PMD) effects. All linear distortion due to PMD and CD can be equalized in the DSP domain, employing algorithms in either offline processing or real-time.

It is thus very important to ensure that these sub-systems are performing the coherent detection and transmitting functions. This section thus presents a summary of the tests conducted with a back-to-back and transmission of QPSK PDM channels. The symbol rate of the transmission system is 28 GBd under the modulation format PDM-CSRZ-QPSK. We note that the differential QPSK (DQPSK) encoder and the bit pattern generator are provided in the optical transmitter. A QAM modulator using $LiNbO_3$ integrated optics and two planar lightwave circuits (e.g. silicon nitrate) is shown in figure 3.5. The electrical

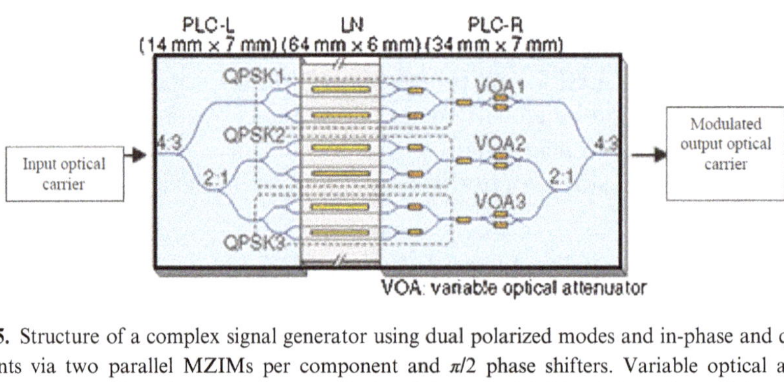

Figure 3.5. Structure of a complex signal generator using dual polarized modes and in-phase and quadrature components via two parallel MZIMs per component and $\pi/2$ phase shifters. Variable optical attenuators (VOAs) are used to adjust the quadrature amplitude of the optical fields. PLC = planar lightwave circuit; L = left; R = right; LN = lithium niobate ($LiNbO_3$). See also the schematics given in figures 3.7 and 3.8.

broadband signals are applied to the electrodes of the MZIMs. Two polarized optical channels are guided through sections of MZIM so that the in-phase and quadrature components are generated and then summed up at the output. Two PLC sections are connected at interfaces of the $LiNbO_3$ so that the phase shifts via thermal effects and variation of the amplitude attenuation. PLC Si_3N_4 of low propagation loss is used to minimize the total insertion loss. Lightwaves modulated by this QAM modulator reserve their coherence and are commonly employed in high speed broadband coherent and intensity modulated transmission systems. The analog signals are fed into an RF amplifier driver and then fed into the electrodes of the MZIMs. Digital waveforms are generated in a DSP system of either frequency domain, e.g. OFDM, or time domain multi-level modulation format, e.g. PAM-*n*, complex M-ary QAM. Pre-distortion or pre-emphasis of the signals can be implemented in this DSP-based transmitter so as to compensate for the transmission fiber lines or any band limited sub-systems of the transmission systems.

3.3.1 112 G–800 Gbps QPSK coherent transmission systems

It is expected that the coherent polarization multiplexed channel QAM formats with a baud rate of 100 GBd can offer high sensitivity 800 GBps per carrier. 25 GBd × 4 lanes for 100 Gbps modules are quite popular now, but still not high enough for DC networking in the near future and DC access networks. This section gives a brief outline of the conceptual design of an optical transmitter and receiver for a 400 Gbps bit rate whose spectrum would look like the 28 Gb s^{-1} symbol rate.

The spectral distribution of the channel consists of two sub-channels separated by 50 GHz symmetrically allocated from the laser central frequency. The modulation is a multi-level QPSK based on the existing QPSK 100 G structures. The multi-level 16QAM is used in order to increase the capacity by a factor of 2 and thus with the frequency division mux contained within the allowable spectral spacing would allow us to generate and transmit a 400 G capacity optical channel. The generic optical Tx and Rx of these advanced sub-systems are briefly described.

The setting and testing of the currently researched 100 G transmission systems are outlined, in particular the noise contribution under the superposition of the differentially amplified and coherently detected noise on the phase modulated and balanced receiving structure. The detection of the combined constellation would suffer some penalty pending on the Euclidean distance between the constellation points. This is still to be explored. The golden ratio of the 16QAM may offer some advantages. However, we expect at least a 3 dB reduction in the sensitivity of the receiver and thus it may require higher effective number of bits (ENOB) of the DSP sub-system.

The role of digital signal processing in the recovery of received signals is also discussed. This proposed system is discussed for reference.

3.3.2 Modulation format 16QAM

QPSK is considered as a basis/kernel for the multi-level advanced modulation format. Other options are to be considered depending on the noise levels contributed by optical amplifiers and quantum shot noise of the opto-electronic front end. In addition two frequency sub-channels are used whose carriers are generated using an external modulator with suppression of the lightwave carrier.

100 G = 2(DQPSK) × 2(PolMux) × 28 G (the symbol rate). Thus for 400 G = 4 × 100 G—the factor of 4 is obtained by frequency division multiplexing (FDM) and one-order higher level modulation (M-ary QPSK). The frequency multiplexing with the channel frequency carriers is generated by an external optical modulator which is driven by a single microwave sinusoidal wave to create the double sideband carriers, shifted by 25 GHz from both sides of the lightwave carrier. The modulator is based symmetrically on the electrical-optical transfer characteristics of the Mach–Zehnder interferometric modulator at $V_\pi/2$ and the amplitude swing is equal to $V_\pi/2$ so that the carrier can be suppressed. Thus the optical transmission is arranged, as shown in figure 3.6.

3.3.3 Optical modulators

Figure 3.7 shows the existing I–Q modulator employed in the 100 G PolMux optical transmitter for generating 28 G symbols s^{-1} with two orthogonally polarized guided

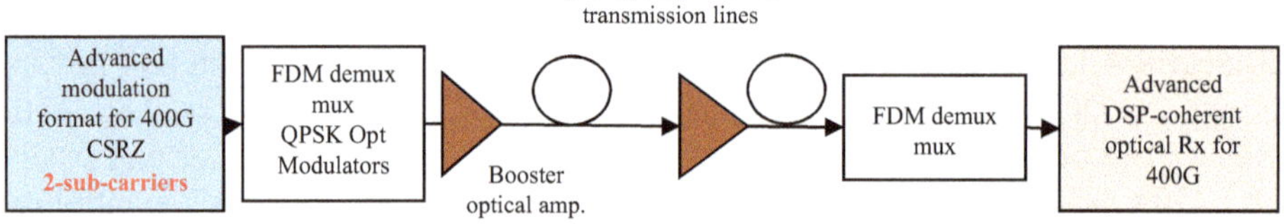

Figure 3.6. 400 G–800 G optical transmission systems.

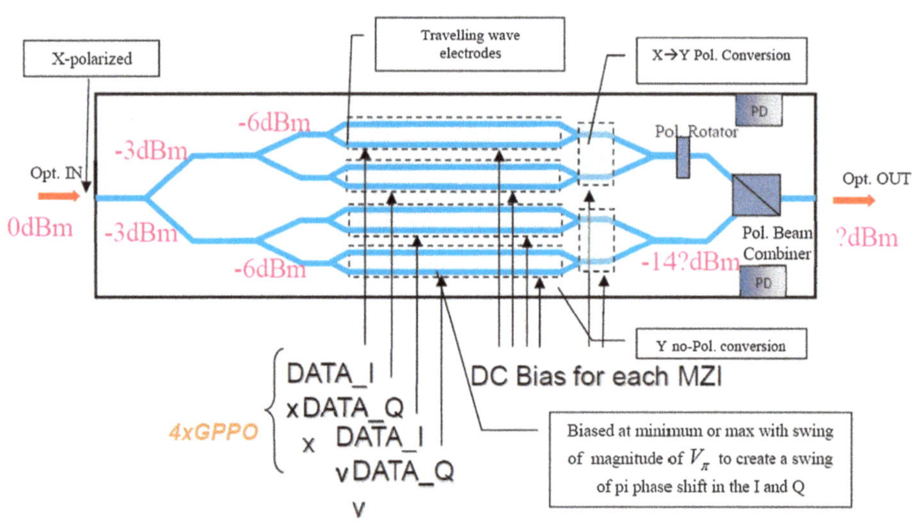

Figure 3.7. Schematic of a Fujitsu I–Q modulator with integrated polarization mux employed in a 100 G Tx coherent transmission system (data specification supplied by Fujitsu). The data sequence applied to the I and Q electrical ports is 28 GS s^{-1}. Note: there is a possibility of biasing point drift due to electro-optic effects generated by static charges accumulated on the surface of the traveling wave electrodes.

lightwave channels. We note the following: (i) the total insertion loss per polarized channel is at least 14 dB; (ii) the modulation is by single electrode driving signals, meaning that the amplitude of the RF electrical signal is quite high, however, this is compensated for by the simplicity of the driving signal requirements and RF connecting; (iii) the biases are implemented by an integrated T-bias built into the optical IC—note that there would be drift of the DC biasing point.

Two carriers can be generated using the CSRZ modulators and then optical filtering, as shown in figure 3.8. However, it is much simpler if two MZIM modulators are used with dual-drive electrodes and the sinusoidal RF waves are $\pi/2$ phase shifted with each other to generate an upper sideband single band and lower sideband instead of using optical filters and one CSRZ optical modulator. This structure is shown in figure 3.9. It is quite straightforward that when the two RF sinusoidal waves are shifted by $\pi/2$ then the lower sidebands cancel each other, resulting in one sideband in the optical domain (figure 3.10), and similarly for the lower sideband generation. The frequency spectra at different positions of the optical FDM modulators are shown in figure 3.11.

It is noted that if this FDM and dual constellation multiplexing are used for a 100 G transmitter then the symbol (baud) rate can be halved, thus the dispersion tolerance and hence transmission distance can be quadrupled. It is assumed that the

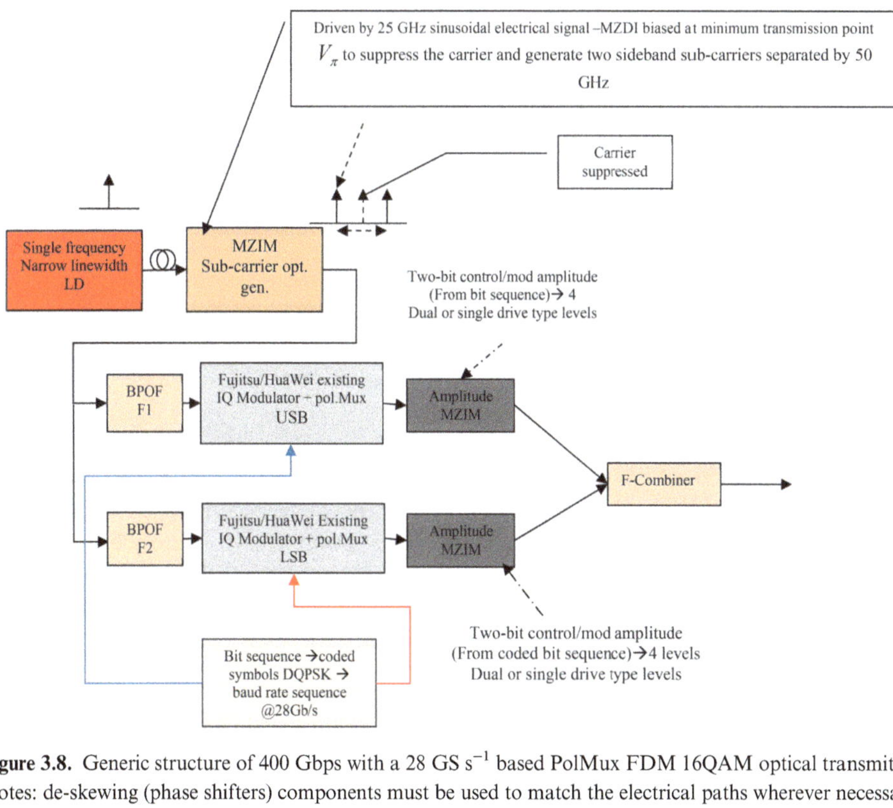

Figure 3.8. Generic structure of 400 Gbps with a 28 GS s^{-1} based PolMux FDM 16QAM optical transmitter. Notes: de-skewing (phase shifters) components must be used to match the electrical paths wherever necessary.

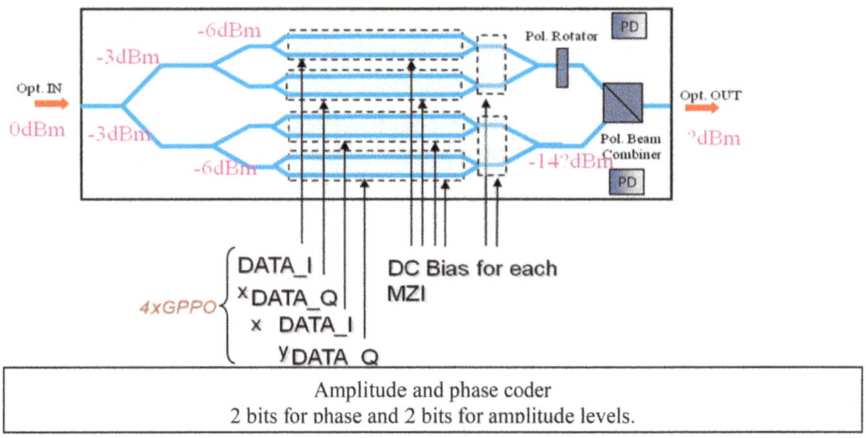

Figure 3.9. Structure of the optical transmitter using four-level APSK with amplitude modulation and the Fujitsu I–Q PDM modulator shifted.

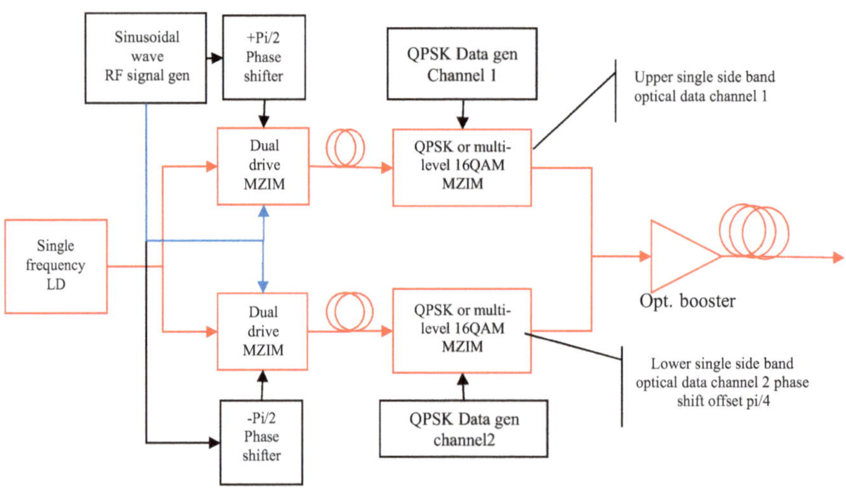

Figure 3.10. Schematic diagram of the optical transmitter using 2 × FDM (upper and lower single sideband) and $\pi/4$ phase shift between the QPSK constellation.

digital optical coherent receiver is the same and that the detection performance is the same for both QPSK and DQPSK. At worst the penalty for 8PSK (multiplexed constellation) could be about 3 dB for the 8PSK compared to the QPSK case.

Using the multiplexed constellation optical coherent receiver, we can simplify the receiver configuration using only one receiver rather than two digital optical receivers for each constellation.

3.3.4 Optical transmitter

The structure of the optical transmitter is shown in figure 3.8, except that the optical band pass filters are not used and only one channel is employed. Figure 3.8 shows a

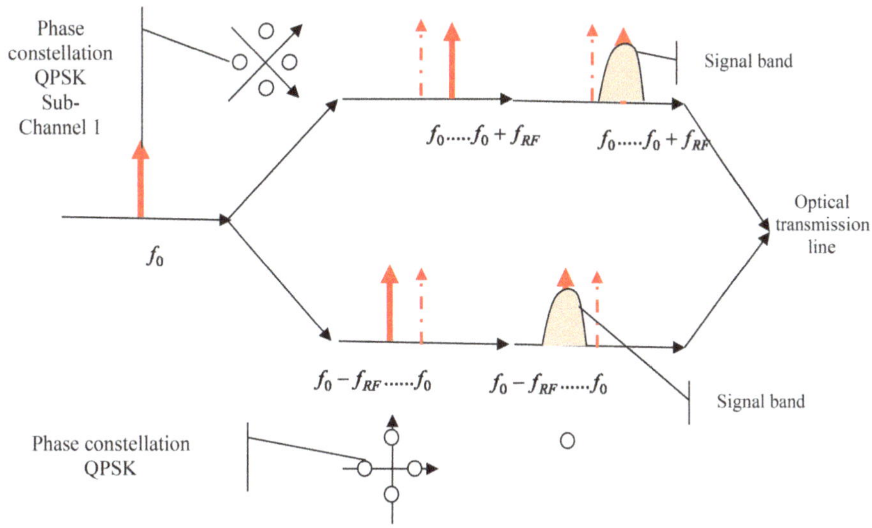

Figure 3.11. Optical frequency spectrum and constellation at different positions of the transmitter described in figure 3.10.

400 Gb s^{-1} optical transmitter. The modulation scheme is DQPSK with polarization multiplexing and de-multiplexing with forward error coding (FEC). Thus for 100 Gb s^{-1}, this modulation scheme would result in 28 GS s^{-1} (baud/symbol rate).

The CW lightwave is generated by a narrow linewidth laser and fed into a Mach–Zehnder amplitude modulator biased at the minimum transmission point of the electrical-optical transfer curve so that a sinusoidal electrical signal at a frequency equal to half of the bit rate and a swing voltage of 2 Vπ would give an optical clock sequence at the output. This is the pulse carver. This optical clock pulse sequence would have the original optical carrier suppressed.

The Fujitsu optical modulator shown in figure 3.7 is used to modulate and generate two polarized lightwave channels. Each polarized optical path is fed into an I–Q optical modulation section in which two parallel MZIM sub-sections of $\pi/2$ optically delayed with each other are employed, accepting the electrical RF data sequence inputs to modulate the phase of the optical carriers.

Following the modulation, the polarized channels are then combined and launched into the transmission fiber line.

Note that if 400 GB s^{-1} is to be designed then another factor of 4 can be achieved by 16 amplitude differential quadrature phase shift keying (ADQPSK) and frequency division multiplexing. A proposed structure for this transmitter is shown later in this chapter. The multi-level modulation format of order N would give another factor to reduce the bit rate to the base baud rate of 28 GS s^{-1}. In this case, $N = 16$ gives another factor of 2 higher than that of the DQPSK scheme. Thus with the FDM and the 4 bits/symbol of the high level modulation scheme, 400 G would give a base baud rate of 28 GS s^{-1} transmission sequence.

3.3.5 Optical receivers

3.3.5.1 Remarks

The receiver consists of: (i) an FDM demultiplexer for each 400 G channel and an array waveguide grating (AWG) if DWDM operation is employed; (ii) a polarization demux to separate two polarized lightwave channels and demodulator for multi-level detection, including intensity detection and DQPSK demodulators at 100 G (a Hua Wei 100 G receiver); and (iii) a digital processing unit for decision making of the phase states of the received I and Q channels.

Obviously, the polarized channels are split into the X- and Y-polarizations and then fed into a wavelength demultiplexer so that each wavelength channel can be filtered. Then each wavelength channel can be fed into optical filters, a demultiplexer or a reconfigurable add/multiplexer (ROADM) to extract channel(s) for destinations or continuing on to further transport.

The optical receiver can use either direct detection or coherent detection. Both these reception techniques have been extensively developed over the years, their difference is based on whether or not the incoming signals are mixed with a high power local oscillator (LO) laser, hence the detection of the beating signals in the optical field rather than intensity and hence current in the photodetector.

In both cases, digital signal processing techniques are used in order to exploit the rich knowledge of processing algorithms to overcome the many difficulties faced by classical non-DSP reception, such as advanced forward error coding, feed forward equalization, recovery of the clock rates, etc. These techniques are explained in more detail in sections 3.6–3.8.

3.3.5.2 Basic analyses of optical receiver noise

At the receiver the optical differential signals are fed into a balanced receiver which consists of two photodetectors (PDs) connected back-to-back (B2B) as differential inputs into an electronic differential amplifier with a differential optical-voltage transfer gain of 2800 V W^{-1}. This differential amplifier then amplifies the differential current and converts the RF output and its complementary to electrical voltage output. The total equivalent noise density is 80 pA/$\sqrt{\text{Hz}}$ and a 3 dB bandwidth typically specified at 30 GHz. This gives a total equivalent noise current $i_{\text{Neq.}}$ of

$$i_{\text{Neq.}} = \sqrt{S_{\text{Neq.}}B} = \sqrt{80^2.\ 10^{-24} \times 30.10^9} = 8 \times \sqrt{3} \times 10^{-6}\text{A} = 8 \times \sqrt{3}\ \mu\text{A}. \quad (3.1)$$

This is the total current represented at the input of the differential amplifier. If this total equivalent noise is brought forward to the input of the amplifier then this amplifier is now considered to be noiseless.

Superimposing on top of this total equivalent electronic noise current, there are a number of noise sources present at the input of the electronic differential amplifier, as follows.

Quantum shot noise is generated by the large power local oscillator mixing with the signal optical power. Assuming that this LO is about 10 dB above the signal power level, then this quantum shot noise dominates that of the signal.

The signal dependent shot noise is thus considered to be much smaller than that of the LO quantum shot noise. Thus the total noise superimposed on top of the signal level is the total current formed by both the quantum shot noise of the LO and the total equivalent electronic noise of the differential amplifier.

The signal current and hence the output differential voltage level can be estimated by considering the differential power of state |0| and state |1| or 0 and π radians of the phase states of the modulated symbols. For example, if an average power of −10 dBm signal power is launched into the U2T receiver then the differential amplitude level is −0.1 mW and +0.1 mW or the total differential power is 0.2 mW. This would give a differential output voltage level of 2800 V W^{-1} × 0.2 mW = 4.8 mV.[1]

Thus, if the power of the LO laser is 13 dBm fed into the 90° hybrid coupler whose total insertion loss is 13 dB, then the LO power is 0 dBm per polarized channel which generates a quantum shot noise of

$$i_{\mathrm{NLO}} = \sqrt{2qP_{\mathrm{LO}}\Re B} = (2 \times 1.6 \times 10^{-19} \times 10^{-3} \times 1.30 \times 10^{9})^{1/2} \approx 1 \times 10^{-6} = 1\ \mu\mathrm{A} \quad (3.2)$$

This current is about nine times less than that of the total equivalent electronic noise, as seen at the input port of the differential amplifier. For a differential optical power of 0.2 mW and a transfer coefficient of the differential amplifier of 2800 V W^{-1}, we would expect an output voltage level of 5.6 mV. Alternatively, the equivalent signal current is 0.75 A W^{-1} which gives a total signal differential current of 0.3 mA. Thus the signal-to-noise ratio (SNR) can be estimated to be

$$\mathrm{SNR} = \frac{\langle i_s \rangle^2}{i_{\mathrm{NLO}}^2 + i_{\mathrm{Neq}}^2} = \frac{300}{1 + 8 \times 1.732} = 20 \text{ or } \mathrm{SNR} = 13\ \mathrm{dB}. \quad (3.3)$$

This is sufficient for optical receivers to operate successfully. Note that the receiver sensitivity is specified at −10 dBm. Thus the LO level boosts the signal power to the appropriate level for the receiver to generate an optical pulse sequence for digital sampling and processing.

3.3.5.3 ENOB and clipping effects of real ADC and AGC on the transmission performance of a coherent QAM system

The roles of the ADC ENOB and AGC clipping effects are investigated in a coherent QAM optical transmission system using experiments and simulations under different transmission effects. Potential implementation risks and optimization for AGC and ADC are presented in this section.

3.3.5.3.1 ENOB and optical systems

Since the development of 100 Gbps optical transmission systems, coherent detection and digital signal processing (DSP) technologies seem to be indispensable as well for systems

[1] Technical specification of the U2T 43 Gb/s DPSK balanced optical receiver. Product Code: BPRV2125(A), www.U2T.com.

beyond 100 G [5–9]. In a coherent system, an analog-to-digital converter (ADC) plays a special role, converting data from the analog into the digital domain. There are two critical ADC issues: the effective number of bits (ENOB) and the clipping effect. The requirement for ADC bit resolution in coherent systems is discussed in [6]. A more realistic description of ADC noise assumes a Gaussian distribution of electrical noise rather than the uniform distribution of noise generated by the quantization process [10]. An automatic gain controller (AGC) implemented in front of the ADC should be considered and optimized together with the ADC. The AGC output signal strength is monitored and compared to the reference voltage (RV) that may be fixed or derived from some optimization algorithms. In addition to clipping, a nonlinear AGC gain transfer function introduces additional system penalties. It becomes more critical for higher modulation formats, e.g. 16QAM systems, since it requires higher ADC resolution.

In transmission systems with different channel impairments the AGC optimization and required ENOB should be analyzed in detail [11]. They strongly depend on the amount of channel distortion and the considered transmission scenario. Assuming realistic ADC and AGC behavior, we studied the ENOB and the clipping effect in the coherent 112 G PDM-DQPSK and 224 G PDM-16QAM optical transmission systems using experiments and simulations. The required optical signal-to-noise ratio (OSNR) at a bit error ratio (BER) of 10^{-3} was estimated over the ENOB and RV ranges for real transmission systems and simulated systems including linear effects: (i) first-order polarization mode dispersion (PMD), which is characterized by the state of polarization angle (SOPA) and differential group delay (DGD), and (ii) chromatic dispersion (CD).

3.3.5.3.2 Experimental and simulation configuration

The 112 Gb s^{-1} PDM-DQPSK was experimentally investigated in a system with a channel spacing of 50 GHz and a span loss of 22 dB. In the DWDM transmission, eight channels of 112 G PDM-DQPSK signals were transmitted over a 1500 km link consisting of 20 spans. Each span consists of a 75 km standard single-mode fiber (SSMF) and a two-stage erbium-doped fiber amplifier (EDFA), in which a section of dispersion compensation fiber (DCF) can be inserted. In the hybrid transmission of 112 G and 40 G, one 112 G channel was set in the middle of eight 40 G QPSK channels. In this case, the CD was fully compensated using DCFs. In the hybrid transmission of 112 G and 10 G, the scenario was the same except that the link length was 900 km. In all cases, the 112 G signal was detected by a coherent receiver that includes a conventional polarization diversity 90 degree hybrid together with the local laser and balanced photo diodes outputting four electrical signals. These electrical signals were sampled by a real-time oscilloscope with 50 G samples/symbol (GSa/sy) and 20 GHz bandwidth. The offline data were then processed by the DSP program based on the Matlab platform, including re-sampling, CD compensation, clock recovery, PMD compensation, carrier frequency compensation, carrier phase recovery, decision and BER estimation (figure 3.12).

The AGC and ADC models were implemented before data re-sampling. A realistic ADC with an AGC can be modeled by two ideal ADCs, an additive noise source and an AGC amplifier [12]. Assuming the Gaussian distribution of electrical

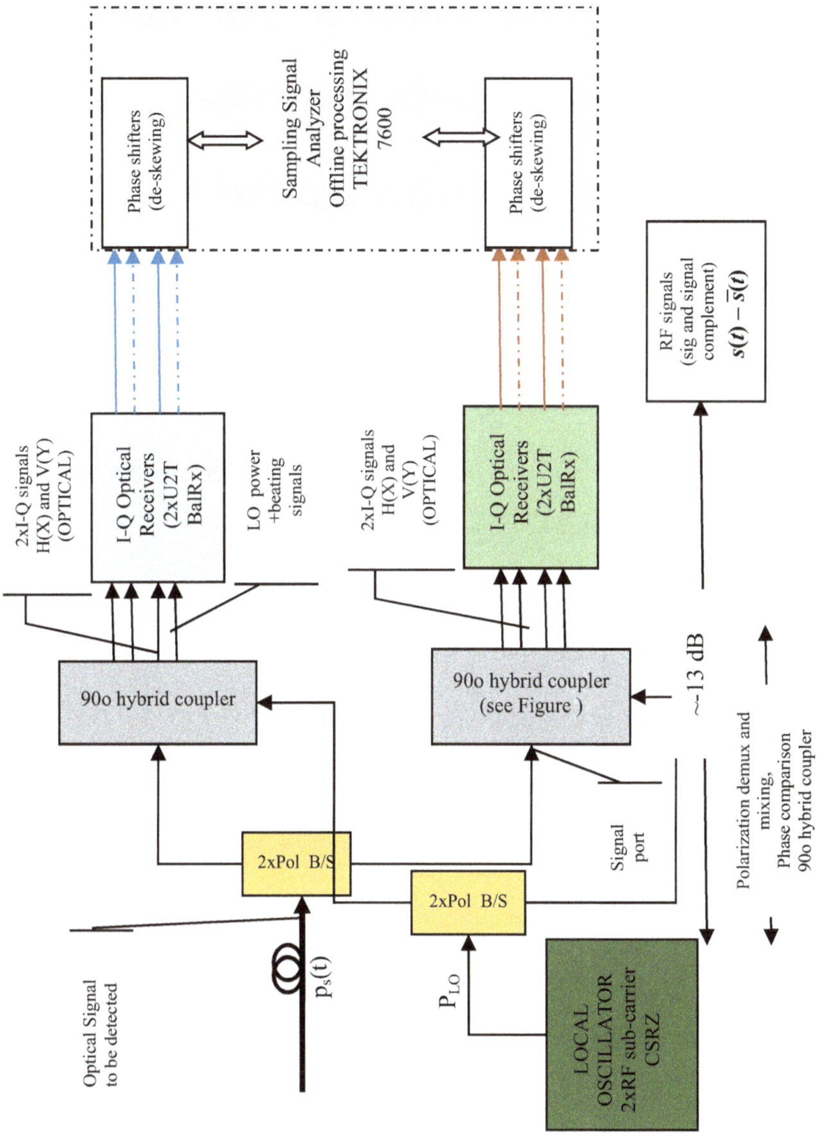

Figure 3.12. Coherent optical receivers including polarization de-multiplexing.

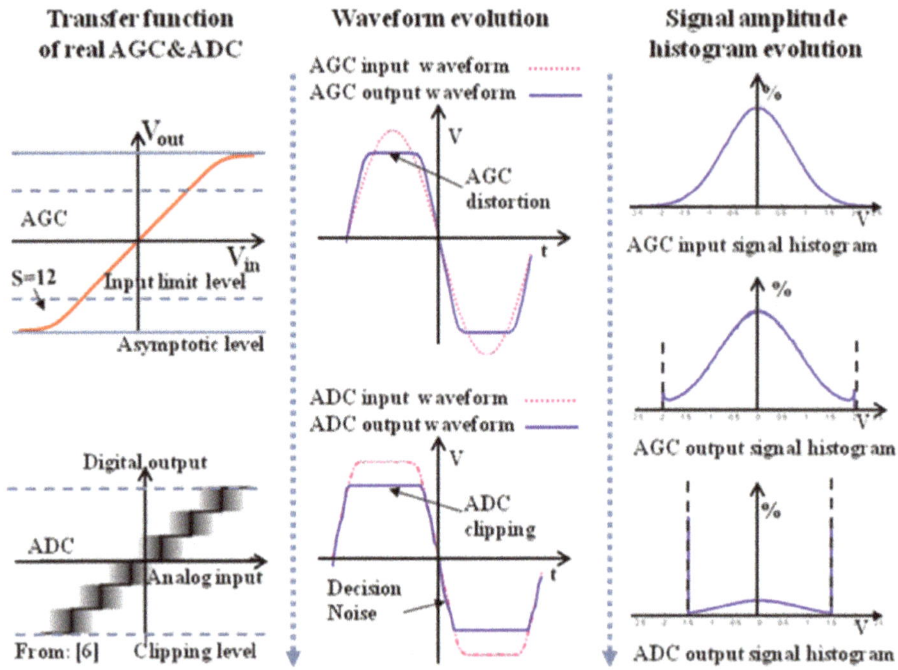

Figure 3.13. Demonstration of the signal waveform and amplitude histogram evolution with the transfer functions of real AGC and ADC.

noise presented at the ADC decision levels, the ADC was modeled with variable ENOB. Regarding the AGC model, the absolute AGC input limit level was set to 1.5 (clipping starts at this point), the absolute asymptotic level was set to 2 (clipped signal within ±2) and the knee sharpness parameter was set to 12 [13]. The 8 bit ADC input limit was set to 1.5, which corresponds to the AGC linear region. Figure 3.13 shows the transfer functions of realistic AGC and ADC along with the time domain waveforms and amplitude histograms of signals before and after these circuits. Large clipping can be observed in the outer ADC bins when AGC gain is set to high value.

3.3.5.3.3 Simulation and experimental results

Simulation results for the DQPSK modulation format are presented in figure 3.14(a)–(c). As expected, channels with CD and PMD are more sensitive to the ENOB value. Even with the appropriate selection of the RV value, the ADC with ENOB = 3 bits results in very bad performance. With an ENO of 3 and under the B2B case, DQPSK may achieve acceptable performance with an optimum performance AGC. However, under channel distortions, the performance is decreased by almost 1 dB. At higher ENOB values the system performance is less sensitive to the RV parameter. Low AGC gain can be compensated by finer ADC resolution. However, when AGC gain is too high the ADC input signal is clipped, and it cannot be compensated by finer ADC resolution. These limitations rapidly

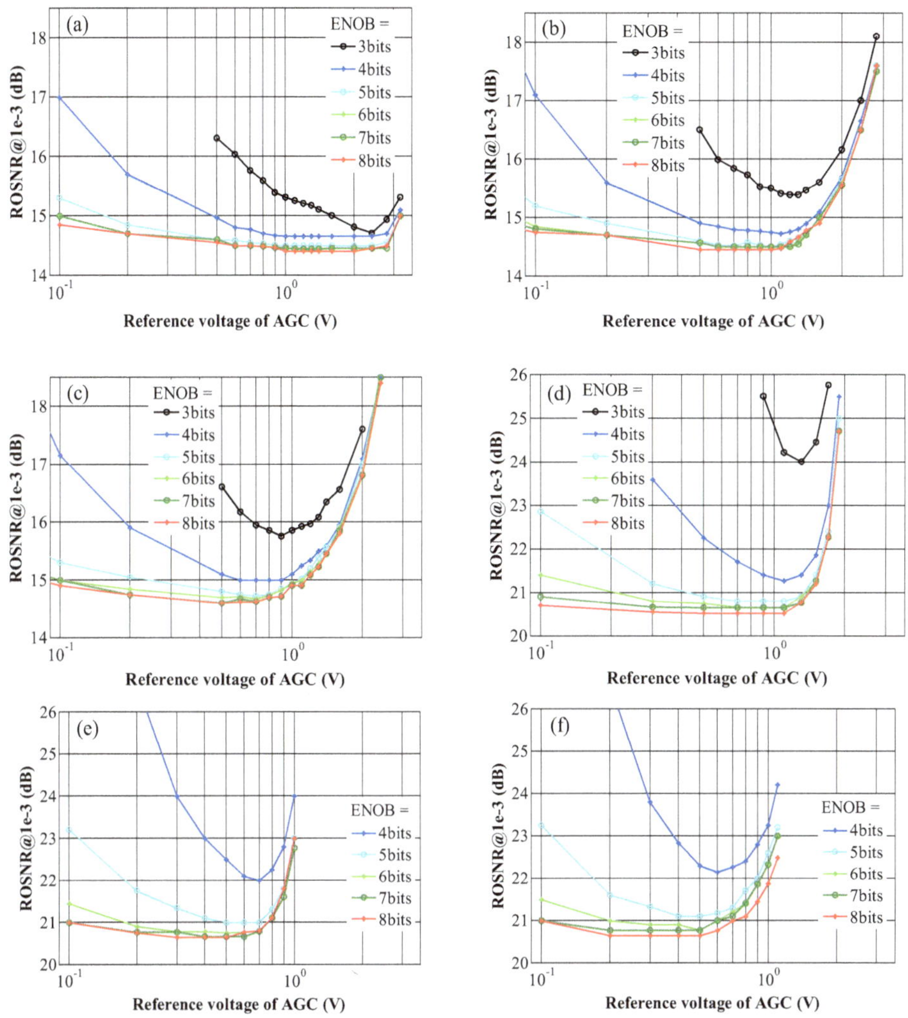

Figure 3.14. Simulation results of the 112 G PDM-DQPSK system: (a) B2B; (b) DGD = 30 ps; SOPA = 45°; (c) CD = 10 000 ps nm^{-1}; DGD = 30 ps; SOPA = 45°. Simulation results of the 224 G PDM-16QAM system: (d) B2B; (e) DGD = 30 ps; SOPA = 45°; (f) CD = 10 000 ps nm^{-1}; DGD = 30 ps; SOPA = 45°.

increase with RV value. In contrast to distorted channels, the B2B can accept a larger RV range. An ENOB value of 5 provides good DQPSK performance although carefully optimized AGC enables acceptable performance (0.2 dB loss) with ENOB = 4. The most critical case for ADC resolution is the channel with high CD. The optimum RV value also depends on the transmission impairments. In the B2B case this sensitivity is not emphasized. However, when distortions are present the optimum value depends on the ENOB. With the current ADC technology, the ENOB is close to 5 bits. Looking at simulation results one can observe that the optimum RV value varies between 0.5 and 1.

Similar effects are observed with 16QAM systems, presented in figure 3.14(d)–(f), with the exception that an ENOB of 3 generates unacceptable penalties. Compared to DQPK one bit more is required to approach the optimum working regime. The AGC clipping degrades the performance for high values of the RV parameter. The AGC dynamic range is more limited by low ENOB than in the DQPSK case. Also, the optimal RV value is much smaller in 16QAM systems. Additionally, the optimum AGC gain more strongly depends on transmission impairments. It can vary by more than 50% with and without dispersion. This indicates the need for accurate AGC control based on the monitoring of channel performance and degradation (OSNR, CD, PMD, etc). According to the rapid penalty increase due to the clipping effect at high RV, the possibility of RV drifting to higher voltage levels should be considered and controlled properly.

The results shown in figure 3.14 are confirmed by the experimental results presented in figure 3.15. Since the measured ENOB was close to 5, we were only able to decrease the ENOB in simulations from 5 to 4 and 3. More degradation arises from nonlinear effects. Also, in the B2B case a larger difference between 3 and 4 ENOB likely arises from component imperfections and electrical bandwidth limitations. In any case, for an ENOB of 5 the RV equal to 0.7 provides a good choice for uncontrolled AGC reference values.

In figure 3.15(f) amplitude histograms of some channels are presented for the same RV value. A large difference between B2B and long transmission is observed. In the B2B case two levels pop up while only noise and electrical bandwidth influence the signal. With distortion the histograms become more Gaussian while the signal is corrupted by nonlinear effects, CD and PMD. This is a reason why a lower AGC RV is preferable when distortion is present.

In future optical transmission systems utilizing higher level modulation formats and digital-to-analog converters (DACs) at the transmitter side (OFDM, *n*QAM, *n*PSK), this problem seems to become more critical. In addition to higher sensitivity to ENOB, the AGC nonlinear amplitude function and precise AGC control will be key factors for maximizing the transmission performance.

3.3.5.3.4 Remarks

The limited ENOB and clipping effects of real AGC and ADC cause penalties that can be partly compensated using sophisticated AGC design and transfer function correction. Furthermore, the penalties are enhanced by transmission impairments such as CD, PMD, nonlinear effects, etc. Therefore, the AGC and ADC dynamics must be optimized with appropriate parameters. The problem becomes more serious in future optical networks utilizing tight channel packaging.

3.3.5.4 Quantization errors and noise

ENOB is a very important parameter for the evaluation of a digital processing system, in particular at the ADC and vice versa. Effectively it is the measure of the corruption of the noise levels on the resolution of the useful signal level in terms of the natural log scale. This parameter ENOB should be considered and evaluated for

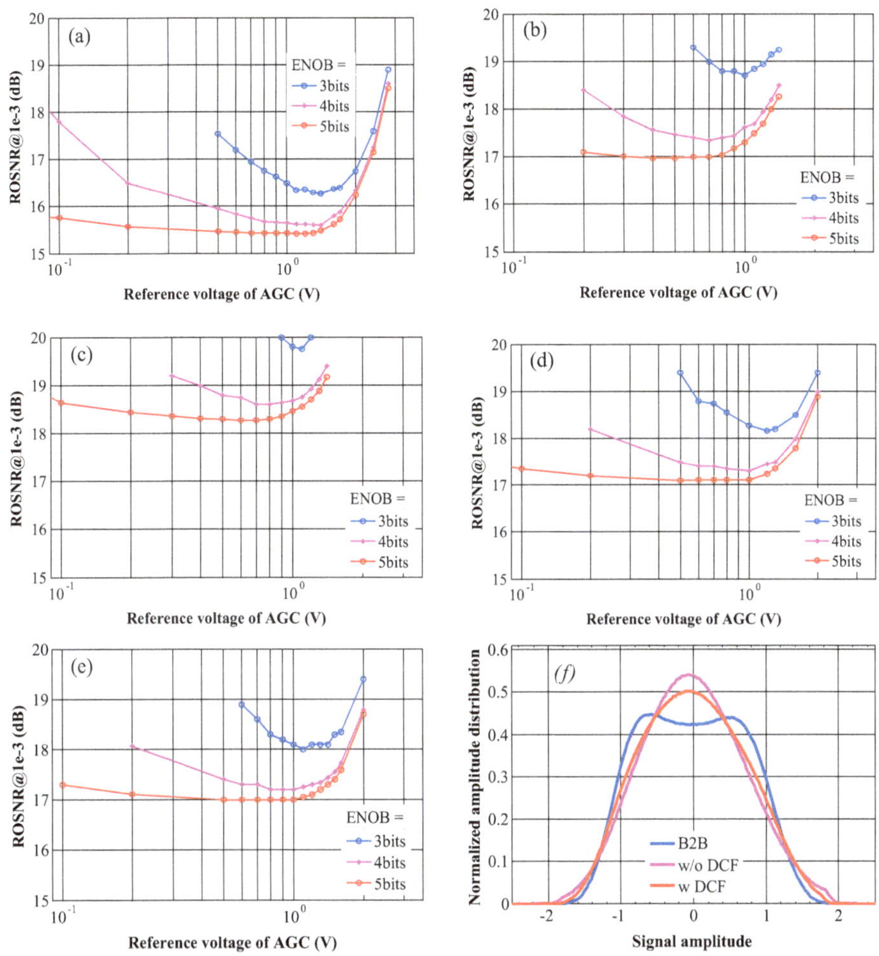

Figure 3.15. Experimental results of 112 G PDM-DQPSKL (a) B2B; (b) DWDM transmission without DCF; (c) DWDM transmission with DCF; (d) hybrid transmission with 40 G; (e) hybrid transmission with 10 G; and (f) signal amplitude histogram before AGC.

digital optical receivers which normally consist of an optical receiver (coherent or non-coherent) followed by ADC and DSP sub-systems. Thus the common definition of the ENOB should now be extended to include the optical receivers at the front end. This optical receiver may include a photodetector or a pair of detectors for balanced receiving techniques and then followed by a wideband electronic amplifier.

The noise of such a digital optical receiver would consist of the following: (i) electronic noise of the ADC and DSP; (ii) electronic noise or equivalent noise spectral density of the electronic wideband amplifier following the photodetectors; (iii) optical noise including accumulated optical amplification noise, signal dependent noise (quantum shot noise) and quantum shot noise due to the high power of the local oscillator if coherent detection is used (see figures 3.16 and 3.17); (iv) nonlinear

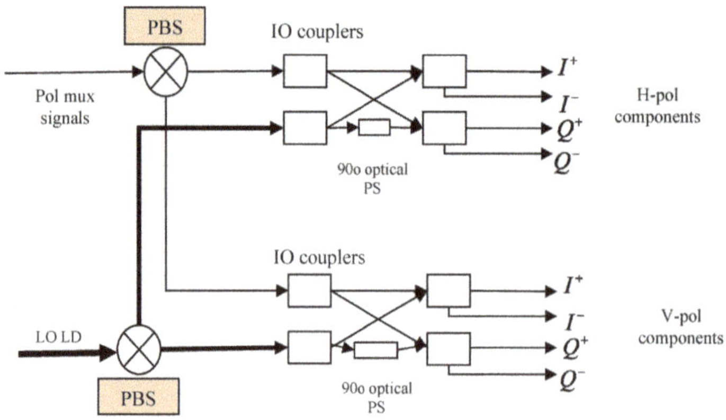

Figure 3.16. 90° hybrid optical coupler with RF output ports of differential difference in the optical phase of the in-phase and quad-phase optical waves. See the differential optical gain conversion factor of the receiver (extracted from an NTT hybrid coupler).

inference or distortion of the data signals, e.g. SPM, XPM and FWM impairments; and (v) linear distortion effects, e.g. chromatic dispersion, PMD, etc.

This noise must be evaluated in terms of the number of the least significant bits (LSB) which would make it ineffective, thus the ENOB of the digital signal processing sub-system. If this is determined to be important then we should conduct a detailed study of this parameter and its impact on receiver performance (figure 3.18).

Furthermore if the two 16APSK constellations are shifted by $\pi/4$ with respect to each other, the homodyne detected constellation would look like those given in figure 3.16. The penalty of OSNR would not be much greater than that of a single 16APSK due to the phase resolution of the $\pi/2$ separation of the 16APSK (figure 3.19).

3.3.6 I–Q imbalance estimation

There are imbalances due to the propagation of the polarized channels and the I- and Q- components. They must be compensated in order to minimize the error. The I–Q imbalance of the Agilent BalRx and U2t BalRx is less than 2 degrees, which might be negligible for the system, as shown in figure 3.20. This imbalance must be compensated for in the DSP domain.

3.4 Optical pre-processing reception and transmitter

The transmission system is arranged as shown in figure 3.21 with the integrated optic pre-processor given in figure 3.2, in association with the coherent reception sub-system. The carrier suppressed return to zero-QPSK (CSRZ-QPSK) transmitter consists of a CSRZ optical modulator which is a single-drive or dual-drive Mach–Zehnder interferometric modulator (MZIM). For the single-drive case, the MZIM is biased at the minimum transmission point of its electrical-power transfer characteristics or depletion of the carrier and driven by a sinusoidal signal whose frequency is half of the symbol rate, e.g. 14 GHz for 28 GBd. A WDM mux is employed to multiplex other wavelength channels located within the C-band (1530–1565 nm). An optical amplifier

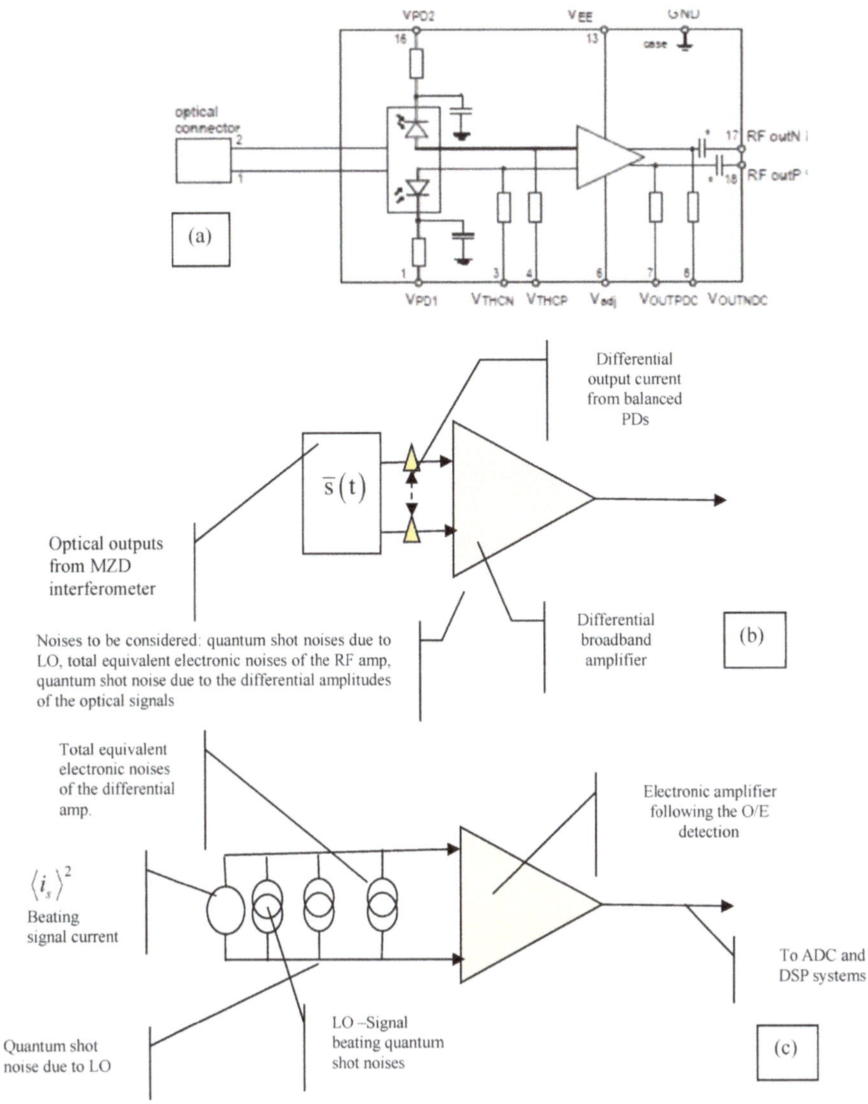

Figure 3.17. Detailed structure of the optical receiver by U2t: (a) general structure and (b) noise representation.

(EDFA type) is employed at the front end of the receiver so that noise can be superimposed on the optical signals to obtain the OSNR. The DSP processed signals in the digital domain are carried out offline and the BER is obtained.

The transmitter lightwave source can be a narrow linewidth external cavity laser (ECL), an Encore type, a polarization splitter coupled with a 45° aligned ECL beam, two separate CSRZ external $LiNbO_3$ modulators and then two I–Q optical modulators. The linewidth of the ECL is specified at about 100 KHz and with external

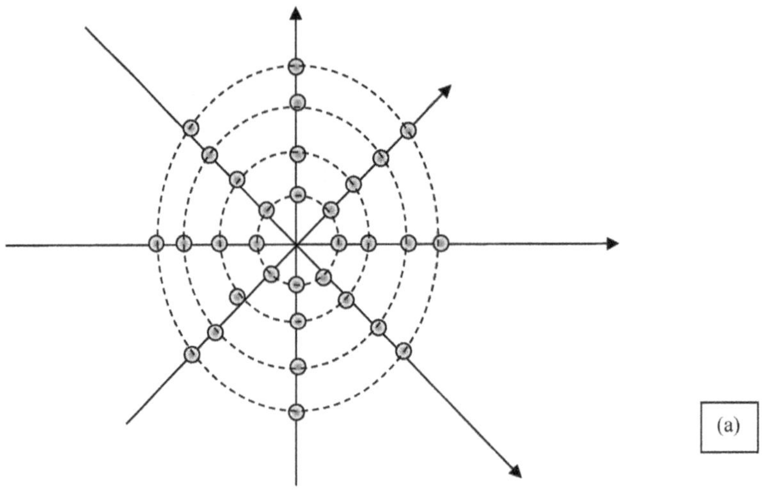

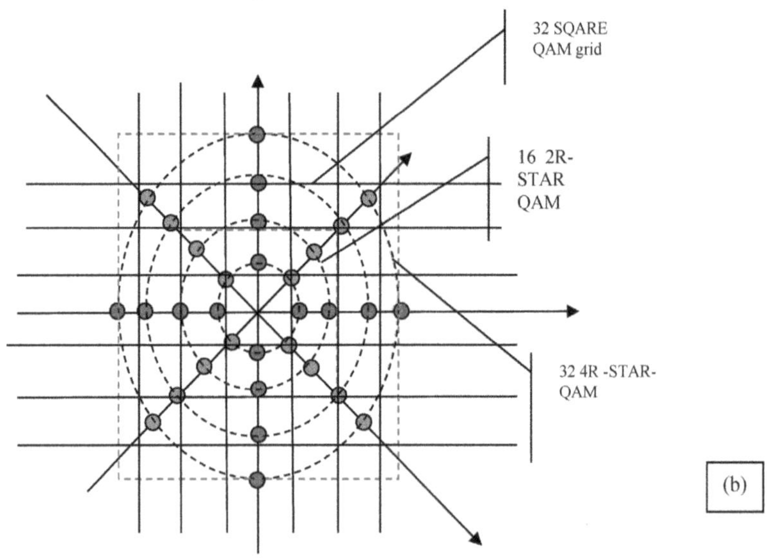

Figure 3.18. Combined 2 × 16APSK to form 4R-32APSK by homodyne detection using two sc 16APSKs: (a) without and (b) with the grid of square QAM.

modulators we can see that the spectrum of the output modulated lightwaves is dominated by the spectrum of the baseband modulated signals. However, we observed that the laser frequency is oscillating about 300 MHz due to the integration of a vibrating grating reflector so as to achieve stability of the optical frequency.

The receivers employed in this system are of two types. One is a commercialized type, Agilent/Keysight N4391 in association with an Agilent/Keysight external local oscillator (LO), the second one consists of a photodetector (PD) balanced pair connected back-to-back in a push–pull manner and then followed by a broadband electronic trans-impedance amplifier (TIA). Further, prior to the electronic reception part, a $\pi/2$ hybrid optical coupler is employed as the optical mixing sub-system

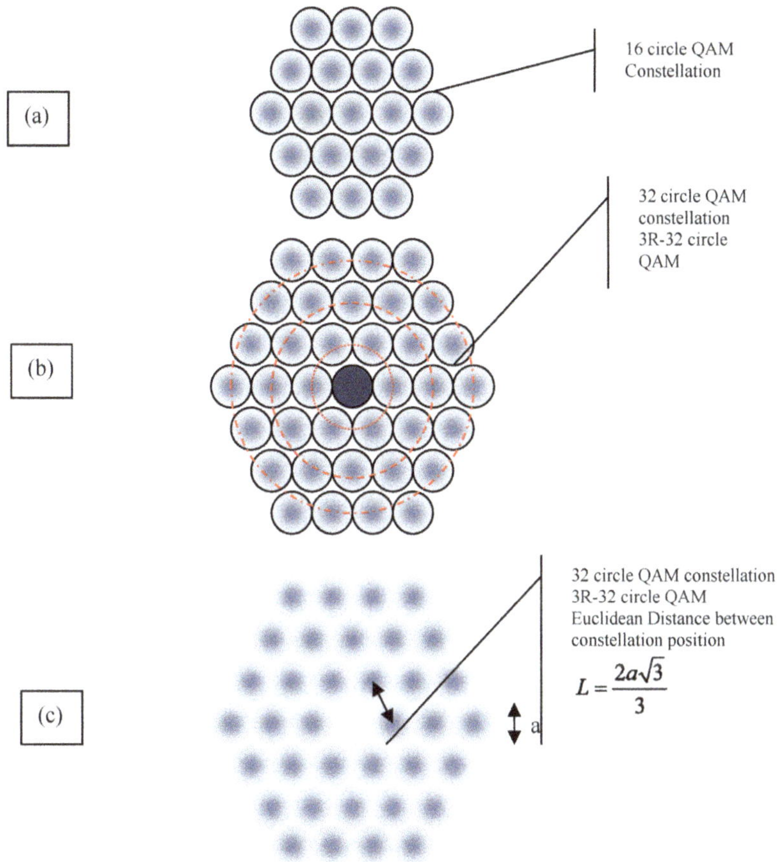

Figure 3.19. Signal constellation of circle QAM: (a) 3R circle QAM; (b) 16 circle QQAM with distance $2a\sqrt{3}/3$; and (c) 3R-32 circle QAM.

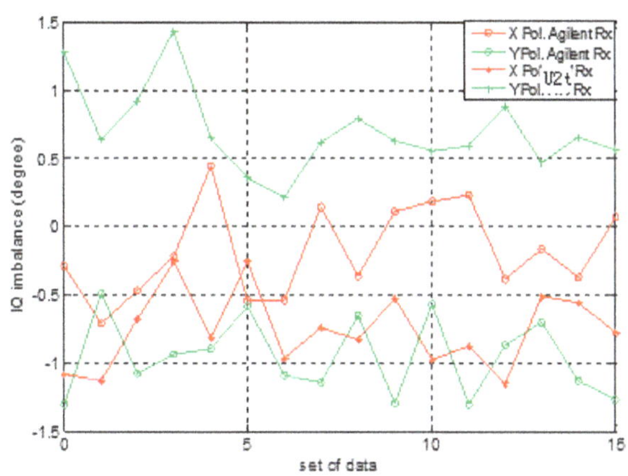

Figure 3.20. I–Q imbalance estimation results for both *Rx*. Note the maximum imbalance phase of ±1.5°.

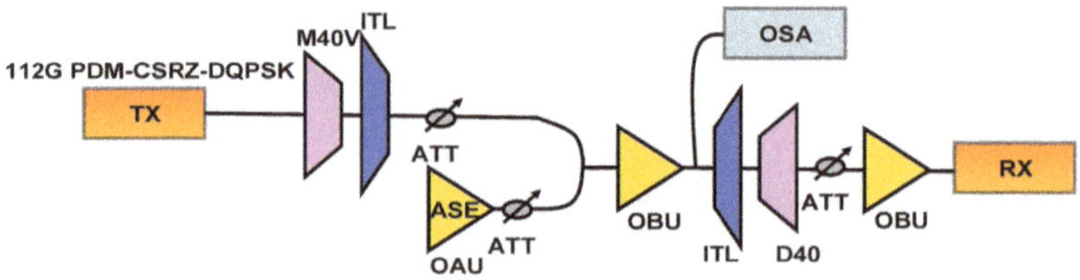

Figure 3.21. Set-up of the PDM-QPSK optical transmission system. Tx = transmitter; PDM = polarization division multiplexing; ITL = interleaved transmission line, return to zero; OSA = optical spectrum analyzer; Rx = receiver; OBU optical boosting unit; D40 = optical demux type 40; Att = attenuator; OAU = optical amplifier unit; ASE = amplification stimulated emission (noise).

at the front end of the receiver. An optical hybrid coupler can be constructed by a polarization splitter, $\pi/2$ phase shift and polarization combiner that mixed the signal polarized beams and those of the LO (an ECL type identical to the one used in the transmitter). The LO mixed polarized beams (I–Q signals in the optical domain) are then detected by balanced receivers. The I–Q signals in the electrical domain are then sampled and stored in the real-time oscilloscope (Keysight DPO). The sampled I–Q signals are then processed offline using the algorithms provided in the scope or one's own developed algorithms such as the evaluation of EVM described above for the Q-factor and hence the BER.

Both the transmitters and receivers are functioning with the required OSNR for the B2B of about 15 dB at a BER of 2×10^{-3}. It is noted that the estimation of the amplitude and phase of the received constellation is quite close to the received signal power and the noise contributed by the balanced receiver with a small difference, due to the contribution of the quantum shot noise contributed by the power of the LO. Figure 3.22 shows the BER versus OSNR for back-to-back QPSK PDM channels. As shown in figure 3.23, for the variation of BER as a function of the signal energy over noise for AWGN noise, for 4QAM or QPSK coherent, then the SNR is expected to be about 8 dB for BER of 1×10^{-3}. Experimental processing of such a scheme in the B2B configuration shows an OSNR of about 15.6 dB. This is due to the 3 dB split by the polarized channels and then additional noise contributed by the receiver, hence about 15.6 dB OSNR is required. The forward error coding (FEC) is set at 1×10^{-3}. The receiver is the Agilent type, as mentioned above.

A brief analysis of the noise at the receiver can be as follows. The noise is dominated by the quantum shot noise generated by the power of the local oscillator, which is at least ten times greater than that of the signals. Thus the quantum shot noise generated at the output of the photodetector (PD) is

$$
\begin{aligned}
i_{\mathrm{N-Lo}}^2 &= 2qI_{\mathrm{Lo}}B \\
I_{\mathrm{Lo}} &= 1.8\ \mathrm{mA}\ @\ 0.9 - \mathrm{PD\_quantum\_efficiency} \\
B &= 31\ \mathrm{GHz} - \mathrm{BW\_U2t\_Agilent\_Rx} \\
i_{\mathrm{N-Lo}}^2 &= 2 \times 10^{-19} \times 1.8 \times 10^{-3} \times 30 \times 10^{9} = 10.8 \times 10^{-12}\ \mathrm{A} \\
\rightarrow i_{\mathrm{N-Lo}} &= 3.286\ \mu\mathrm{A} \rightarrow \mathrm{shot\text{-}noise\text{-}current}
\end{aligned}
$$

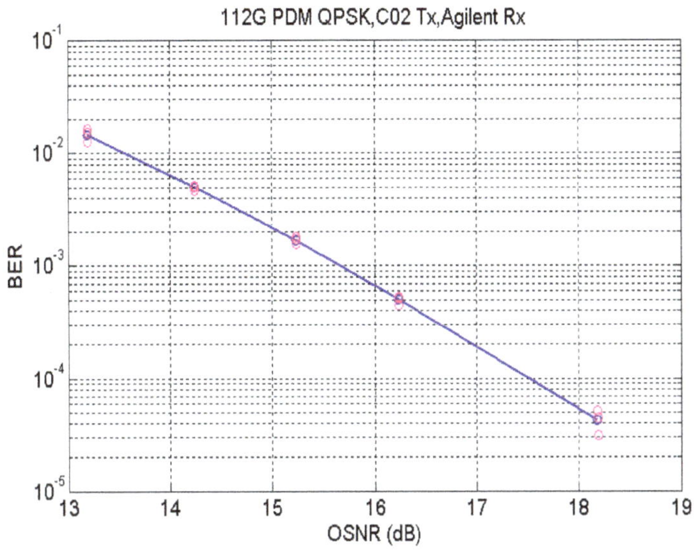

Figure 3.22. Back-to-back OSNR versus BER performance with Keysight Rx in a 112 Gbps PDM-CSRZ-DQPSK system.

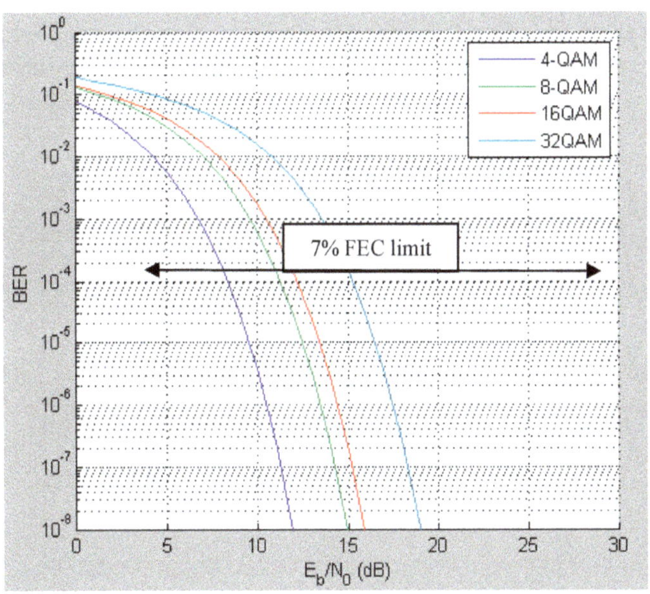

Figure 3.23. Theoretical BER versus SNR for different level QAM schemes obtained using 'bertool.m' of MATLAB.

The bandwidth of the electronic pre-amplifier TIA of 31 GHz is taken into account. This shot noise current due to the LO imposed on the PD pair is approximately compatible with that of the electronic noise of the electronic receiver, given that the noise spectral density equivalent at the input of the electronic amplifier of the U^2t balanced receiver is specified at 80 pA/$\sqrt{\text{Hz}}$, i.e.

$$(80 \times 10^{-12})^2 \times 30.10^9 = 19.2 \times 10^{-12}\text{A}^2 \underset{\text{noise\_current}}{\longrightarrow} i_{\text{Neq.}} = 4.38\ \mu\text{A}.$$

Thus any variation in the LO would affect this shot noise in the receiver. It is thus noted that with the trans-impedance (Z_T) of the electronic pre-amplifier estimated at 150 Ω, a 10 dBm difference in the LO would contribute to a change of the voltage noise level of about 0.9 mV in the signal constellation obtained at the output of the ADC. A further note is that the noise contributed by the electronic front end of the ADC has not been taken into account. We note that differential TIA offers at least ten times higher transfer impedance of around 4500 Ω over the 30 GHz mid-band. These TIAs offer much higher sensitivities compared to the single input TIA type[2], [14].

3.4.1 Skew estimation

In addition to the imbalance of the I and Q due to optical coupling and electronic propagation in high frequency cables, there are propagation delay time differences between these components that must be compensated. The skew estimation is shown in figure 3.24, obtained over a number of data sets.

Abnormal skew variation from time to time was also observed, which should not happen if there is no modification of the hardware. Considering the skew variation that happened with the Agilent receiver, which only has a very short RF cable and

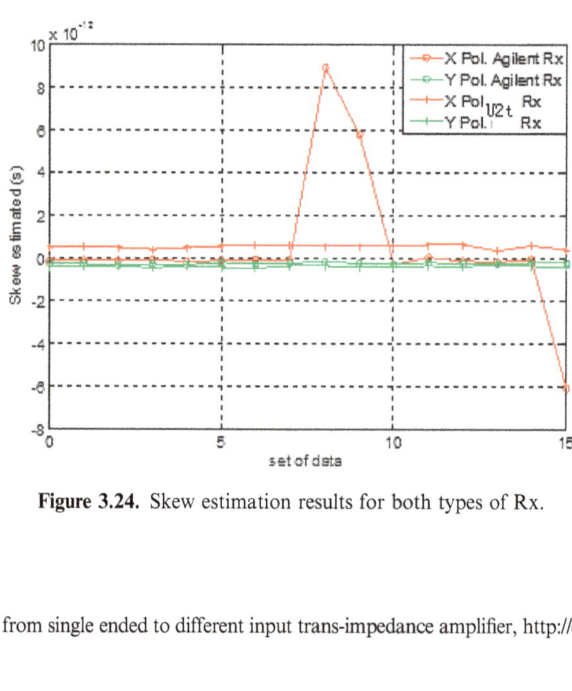

Figure 3.24. Skew estimation results for both types of Rx.

[2] Linear Circuits Inc., from single ended to different input trans-impedance amplifier, http://circuits.linear.com/267.

tight connection, there is a high probability that the skew happened inside the Tektronix oscilloscope more than at the optical or electrical connection outside.

Figure 3.25 shows the BER versus the OSNR when the skew and the imbalance between the I- and Q- components of the QPSK transmitter are compensated. The OSNR of DQPSK is improved by about 0.3~0.4 dB at 1e-3 BER compared to the result without I–Q imbalance and skew compensation as shown in the constellation obtained in figure 3.26. In the time domain, the required OSNR of DQPSK at a BER of 1e-3 is about 14.7 dB, and is improved by 0.1 dB. The required OSNR at 1e-3 BER of QPSK is about 14.7 dB, which is roughly the common performance of the state-of-the-art for QPSK. For a BER = 1e-3 an imbalanced CMRR = −10 dB would create a penalty of 0.2 dB in the OSNR for the Agilent receiver and an improvement of 0.7 dB for a commercial balanced receiver employed in the Rx sub-systems.

Figure 3.27 shows the structure of the external cavity laser incorporating a reflection mirror which is vibrating at a slow frequency of around 300 MHz. A control circuit would be included to indicate the electronic control of the vibration and cooling of the laser so as to achieve stability and elimination of Brillouin scattering effects.

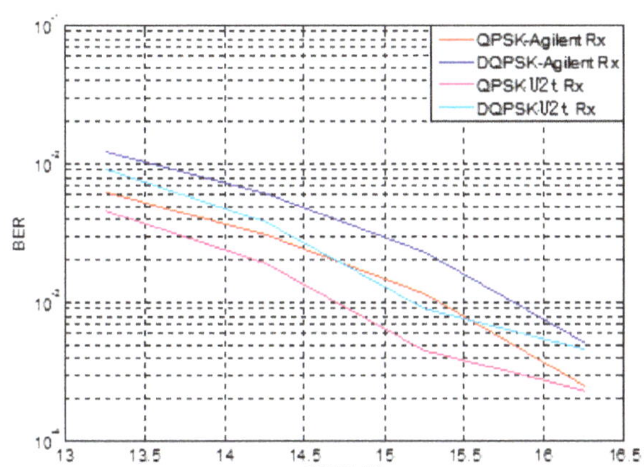

Figure 3.25. OSNR versus BER for two types of integrated coherent receiver after compensating I–Q imbalance and skew.

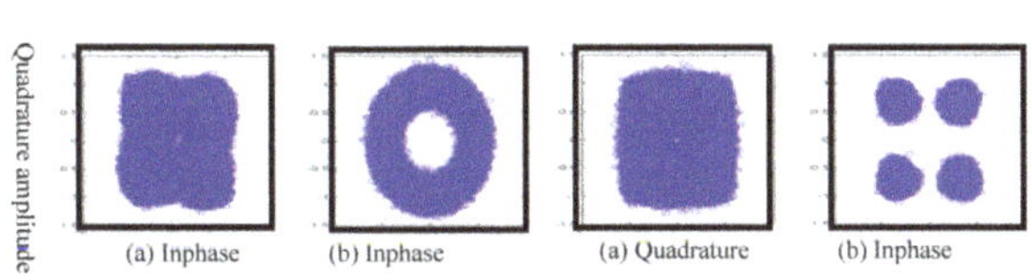

Figure 3.26. Constellation after the PMD module (a) and after the CPE algorithm module (b).

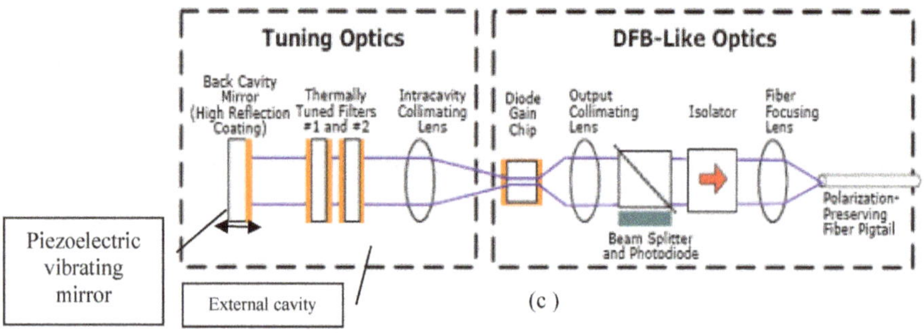

Figure 3.27. Typical noisy QPSK constellation of PDM channels: (a) V-pol and (b) H-pol. (c) External optical cavity structure of a DFB laser with the back mirror vibration used as a narrow linewidth transmitting source. DFB = distributed feedback.

3.4.2 Fractionally spaced equalization of CD and PMD

Ip and Kahn [8] employed the fractional spaced equalization scheme with mean square error (MSE) to evaluate the effectiveness of PDM amplitude shift keying (ASK) with a non-return to zero (NRZ) or return to zero (RZ) pulse shaping transmission system. Their simulation results are displayed in [15] for the maximum allowable CD (normalized in ratio with respect to the dispersion parameter of the well-known standard single-mode fiber (SSMF)) versus the number of equalizer taps N for a 2 dB power penalty at a launched OSNR of 20 dB per symbol for ASK RZ and NRZ pulse shapes, using a Bessel anti-aliasing filter with sampling rate $1/T = M/KT_s$; T_s = symbol_period, with M/K as the fractional ratio.

3.4.3 Electronic digital equalization

Ip and Kahn [15] first developed and applied the back propagation to equalize the distortion due to nonlinear impairment of optical channel transmission through single-mode optical fibers. The back propagation algorithm is simply a reverse phase rotation at the end of each span of the multi-span link. The rotating phase is equivalent to the phase exerted on the signals in the frequency domain with a square of frequency dependence. Thus this back propagation is efficient in the aspect that the whole span can be compensated so as to minimize the numerical processes, hence less processing time and central processing unit time of the digital signal processor is required.

Figure 3.28 shows the equalized constellations of a 21.4 GSy s^{-1} QPSK modulation scheme system after transmission through 25 × 80 km non-DCF spans under the equalization using: (a) linear compensation only, (b) nonlinear equalization and (c) using combined back propagation and linear equalization. Obviously the back propagation contributes to the improvement of the performance of the system.

The phase errors of the constellation states at the receiver can be identified for the launch power of a 25 × 80 km multi-span QPSK 21.4 GSy s^{-1} transmission system. The results of [1] show the performance of back propagation phase rotation per span

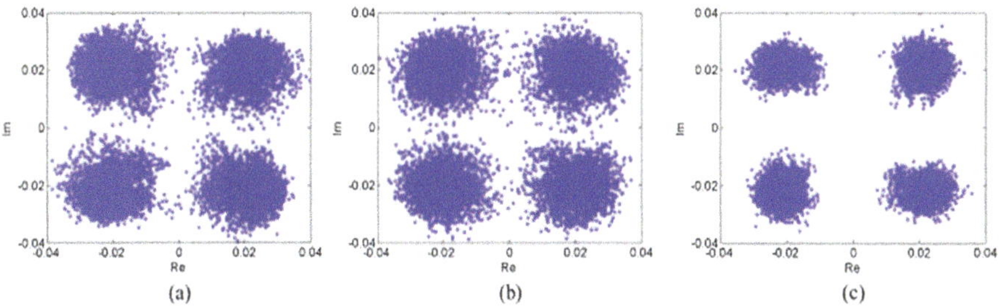

Figure 3.28. Simulated constellation of the QPSK scheme as monitored: (a) with linear CD equalization only, (b) with nonlinear phase noise compensation and (c) after back propagation processing combined with DSP-based linear equalization.

for 21.4 Gb s^{-1} 50% RZ-QPSK transmitted over 25 × 80 km spans of SSMF incorporating five reconfigurable optical add–drop modules (ROADMs), with 10% under compensating chromatic dispersion (CD). The algorithm is processed offline for the received sampled data after 25 × 80 km SSMF propagation via the use of the nonlinear Schrödinger equation (NLSE) and coherent reception technique. It is desired that the higher the launch power the better the OSNR that can be employed for longer distance transmission. So, a fractional space ratio of 3 and 4 offers higher launch power and thus is the preferred equalization scheme compared to equal space or a sampling rate equal to that of the symbol rate. The ROADM is used to equalize the power of the channel under consideration compared to other DWDM channels.

3.5 16QAM systems

Consider the 16QAM received symbol signal whose phase Φ denotes the phase offset. The symbol d_k denotes the magnitude of the QAM symbols and n_k is the noise superimposed on the symbol at the sampled instant. The received symbols can be written as

$$r_k = d_k e^{j\phi} + n_k; \quad k = 1, 2, \ldots.L. \tag{3.4}$$

Using the maximum likelihood sequence estimator (MLSE), the phase of the symbol can be estimated as

$$l(\phi) = \sum_{k=1}^{L} \ln\left\{\sum_{d} e^{-\frac{1}{2\sigma^2}|r_k - de^{j\phi}|^2}\right\}. \tag{3.5}$$

Effectively one would take the summation of the contribution of all states of the 16QAM on the considered symbol measured as the geometrical distance in natural logarithmic scale with the noise contribution of a standard deviation σ.

The frequency offset estimation for 16QAM can be conducted by partitioning the 16QAM constellation into a number of basic QPSK constellations, as shown in figure 3.29. There are two QPSK constellations in the 16QAM whose symbols can be extracted from the received sampled dataset. They are then employed to estimate the phase of the carrier as described in the previous section on carrier phase estimation for

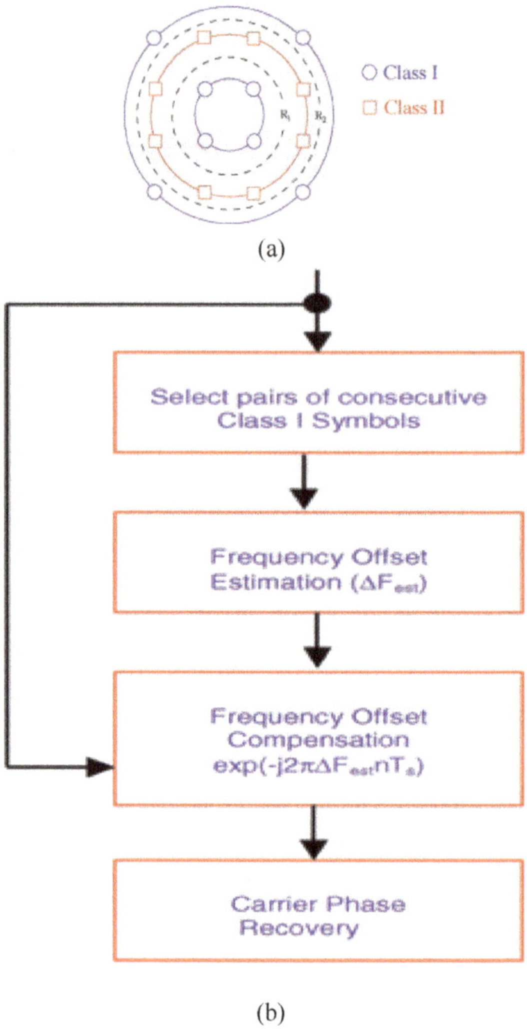

Figure 3.29. Processing of 16QAM for carrier phase estimation: (a) constellation of 16QAM and (b) processing for carrier phase recovery with classes I and II of circulator sub-constellation.

QPSK modulated transmission systems. First, a selection of the inner-most QPSK constellation, classified as class I symbols, is made, then an estimation of the frequency offset of the 16QAM transmitted symbols. Then an FO compensation algorithm is carried out and then the phase recovery of all 16QAM symbols can be derived. Further confirmation of the difference of carrier phase recovery or estimation can be conducted with the constellation of class I, as indicated in figure 3.29(a).

Carrier phase recovery based on the Viterbi–Viterbi algorithm on the class I QPSK sub-constellation of the 16QAM may not be sufficient and so a modified scheme has been reported by Fatadin *et al* in [16]. Further refining this estimation of the carrier phase for 16QAM by partition and rotating are made so as to match

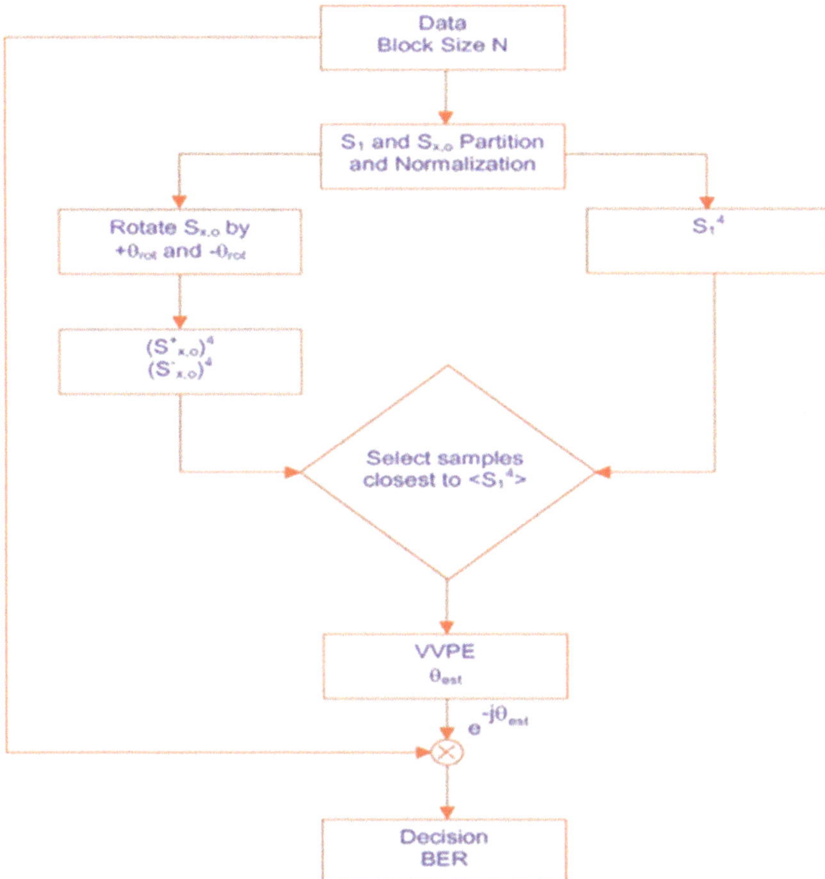

Figure 3.30. Refined carrier phase recovery of 16QAM by rotation of class I and class II sub-constellations.

certain symbol points to those of the class I constellation of the 16QAM. The procedures are as shown in the flow diagram of figure 3.30.

Conduct the partition into different classes of constellation and then rotate class 2 symbols with an angle either clockwise or anticlockwise of $\pm\theta_{\text{rot}} = \pi/4 - \tan^{-1}(1/3)$. In order to avoid opposite rotation with respect to the real direction, the estimation of the error in the rate of change of the phase variation or frequency estimation can be found by using the fourth power of the argument of the angles of two consecutive symbols, given by

$$\Delta F_{\text{est.}} = \frac{1}{8\pi T_s}\arg\left(\sum_{k=0}^{N} S_{k+1}S_k\right)^4, \tag{3.6}$$

to check their quadratic mean, then selecting the closer symbol, then applying the standard Viterbi–Viterbi procedure. Louchet *et al* [17] also employed a similar method and confirmed the effectiveness of such a scheme. The effects on the constellation of the 16QAM due to different physical phenomena are shown in

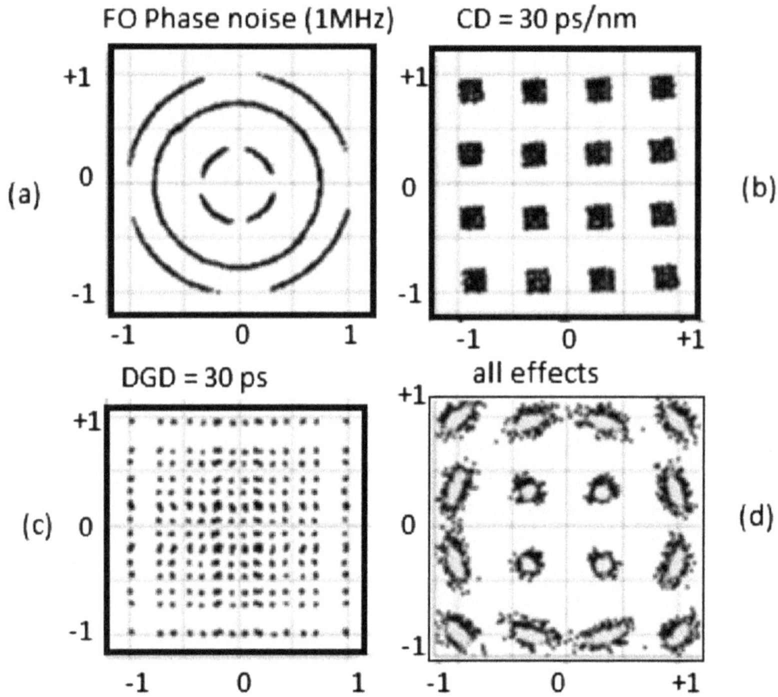

Figure 3.31. 16QAM constellation under influence of (a) phase rotation due to FO of 1 MHz and no amplitude distortion, (b) residual CD impairment, (c) DGD of the PMD effect and (d) total phase noise effect. Adapted from [5].

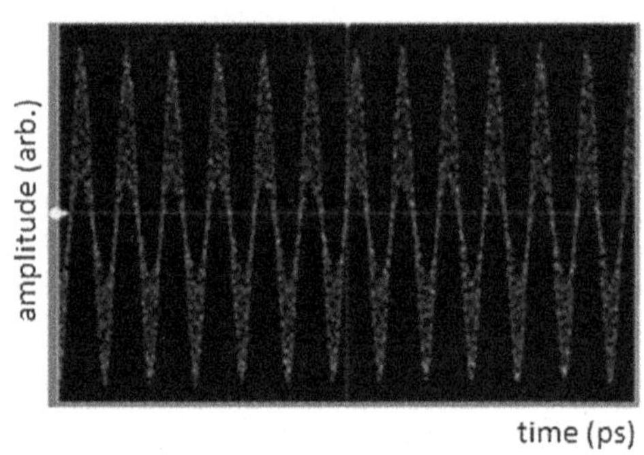

Figure 3.32. Beating signals of the two mixed lasers as observed by a real-time sampling oscilloscope.

figure 3.31. Clearly the FO would generate the phase noise in (a) and influence both the I- and Q- components by CD of a small amount (so as to see the constellation noise) and the delay of the polarized components on the I- and Q- components. These distortions of the constellation allow practical engineers to assess the validity of algorithms, which are normally separate and independent, and implemented in serial mode. This is in contrast to the constellations illustrated for QPSK, as shown in figure 3.26. Figure 3.32 also shows the real-time signals which result from the beating of the two sinusoidal waves of FO beating in a real-time oscilloscope.

Noe *et al* [18] have simulated the carrier phase recovery for QPSK with polarization multiplexed (PDM) channels under CMA and decision-directed (DD) QPSK with and without modification in which the error detected in each stage would be updated. The transmission system under consideration is back-to-back with white Gaussian noises and phase noises superimposition on the signals. For an OSNR of 11 dB the CMA with modification offers a BER of $1 \times 10 - 3$ while it is $4 \times 10 - 2$ for CMA without modification. This indicates that the updating of the matrix coefficients is critical to recover the original data sequence. The modified CMA was also recognized to be valid for 16QAM.

3.6 Terabits/second superchannel transmission systems

3.6.1 Overview

PDM-QPSK has been exploited in 100 Gb s^{-1} long distance transmission commercial systems, and the optimum technologies for 400 GE/1 TE transmission for next-generation optical networking have now attracted significant interest in deploying ultra-high capacity information over the global Internet backbone networks. Tbps transmission systems have also attracted several research groups as the logical rate to increase from 100 Gb s^{-1}. The development of hardware platforms for 1–N Tbps is critical for proving the design concept. The Tbps superchannels can be defined as optical channels comprising a number of sub-rate sub-channels whose spectra would be the narrowest allowable. Thus in order to achieve good spectral efficiency, phase shaping is required and one of the most efficient techniques is Nyquist pulse shaping. Thus Nyquist-QPSK can be considered as the most effective format for the delivery of high spectral efficiency and effective coherent transmission and reception, as well as equalization at both the transmitting and reception ends.

Thus in this section we describe a detailed design and experimental platform for the delivery of Tbps using Nyquist-QPSK at a symbol rate of 28–32 GSa s^{-1} and ten sub-carriers. The generation of sub-carriers has been demonstrated using either re-circulating frequency shifting (RFS) or nonlinear driving of an I–Q modulator to create five sub-carriers per main carrier, thus two main carriers are required. Nyquist pulse shaping is used for effective packing of multiplexed channels whose carriers are generated by the comb generation technique. A DAC with the sampling rate varied from 56–64 GS s^{-1} is used for generating Nyquist shaped pulses, including the equalization of the transfer functions of the DAC and optical modulators.

3.6.2 Nyquist pulse and spectra

The raised cosine filter is an implementation of a low-pass Nyquist filter, i.e. one that has the property of vestigial symmetry. This means that its spectrum exhibits odd symmetry about $1/2T_s$, where T_s is the symbol period. Its frequency domain representation is a 'brick-wall-like' function, given by

$$H(f)=\begin{cases} T_s & |f|\leqslant \frac{1-\beta}{2T_s} \\ \frac{T_s}{2}\left[1+\cos\left(\frac{\pi T_s}{\beta}\left\{|f|-\frac{1-\beta}{2T_s}\right\}\right)\right] & \frac{1-\beta}{2T_s}<|f|\leqslant\frac{1+\beta}{2T_s} \\ 0 & \text{otherwise}\end{cases} \tag{3.7}$$

with $0\leqslant\beta\leqslant 1$.

This frequency response is characterized by two values: β, the roll-off factor, and T_s, the reciprocal of the symbol rate in Sym/s, that is $1/2T_s$ is the half bandwidth of the filter. The impulse response of such a filter can be obtained by analytically taking the inverse Fourier transformation of equation (3.7), in terms of the normalized sinc function, as

$$h(t)=\sin\mathrm{c}\left(\frac{t}{T_s}\right)\frac{\cos\left(\frac{\pi\beta t}{T_S}\right)}{1-\left(2\frac{\pi\beta t}{T_S}\right)^2}, \tag{3.8}$$

where the roll-off factor, β, is a measure of the excess bandwidth of the filter, i.e. the bandwidth occupied beyond the Nyquist bandwidth as from the amplitude at $1/2T$. Figure 3.33 depicts the frequency spectra of the raised cosine pulse with various roll-off factors. The corresponding time domain pulse shapes are given in figure 3.34.

When used to filter a symbol stream, a Nyquist filter has the property of eliminating ISI, as its impulse response is zero at all nT (where n is an integer), except when $n=0$. Therefore, if the transmitted waveform is correctly sampled at the receiver, the original symbol values can be recovered completely. However, in many practical communications systems, a matched filter is used at the receiver, so as to minimize the effects of noise. For zero ISI, the net response of the product of the transmitting and receiving filters must equate to $H(f)$, thus we can write

$$H_R(f)H_T(f)=H(f), \tag{3.9}$$

or alternatively we can rewrite that

$$|H_R(f)|=|H_T(f)|=\sqrt{|H(f)|}. \tag{3.10}$$

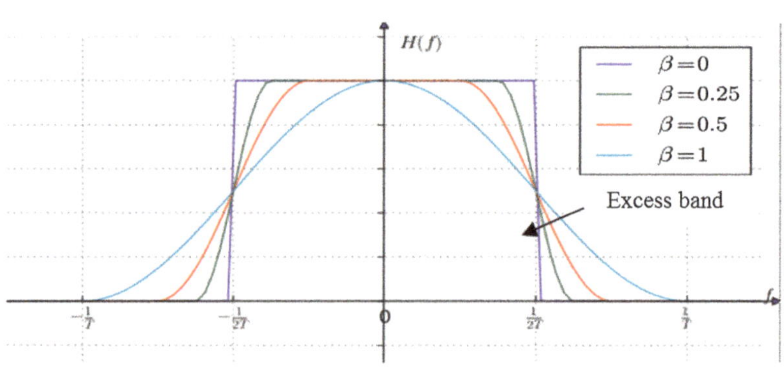

Figure 3.33. Frequency response of the raised cosine filter with various values of the roll-off factor β.

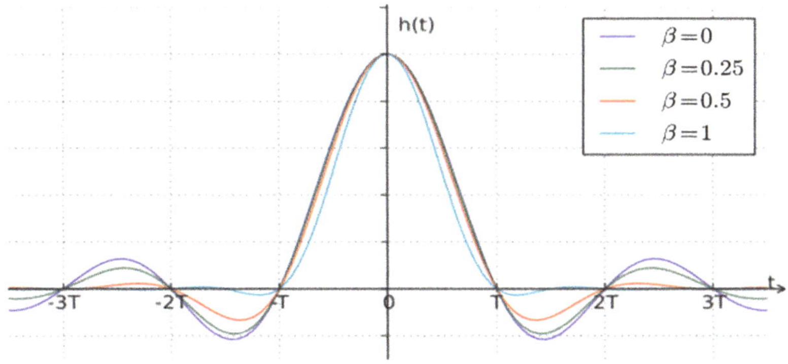

Figure 3.34. Impulse response of the raised cosine filter with the roll-off factor β as a parameter.

The filters which can satisfy the conditions of equation (3.10) are the root-raised cosine (RRC) filters. The main problem with RRC filters is that they occupy a larger frequency band than that of the Nyquist sinc pulse sequence. Thus for the transmission system we can split the overall raised cosine filter with the RRC filter at both the transmitting and receiving ends, provided the system is linear. This linearity is to be specified accordingly. An optical fiber transmission system can be considered to be linear if the total power of all channels is under the nonlinear SPM threshold limit. When it is over this threshold a weakly linear approximation can be used.

The design of a Nyquist filter influences the performance of the overall transmission system. The oversampling factor, the selection of roll-off factor for different modulation formats, and FIR Nyquist filter design are key parameters to be determined. If taking into account the transfer functions of the overall transmission channel including fiber, WSS and the cascade of the transfer functions of all O/E components, the total channel transfer function is more Gaussian-like. To compensate this effect in the Tx-DSP, one would thus need a special Nyquist filter to achieve the overall frequency response equivalent to that of the rectangular or raised cosine with a roll-off factor as shown in figure 3.35.

3.6.3 Superchannel system requirements

Transmission distance. In the next generation of backbone transport, the transmission distance should be comparable to the previous generation, namely the 100 Gbps transmission system. As the most important requirement, we require the 1 Tbps transmission for long distance transmission to be ~1500–2000 km, and for metro application ~300 km.

CD tolerance. As the SSMF fiber CD factor/coefficient 16.8 ps nm^{-1} is the largest among the currently deployed fibers, CD tolerance should be up to 30 000 ps nm^{-1} at the central channel, whose wavelength is approximated at 1550 nm. At the edge of the C-band this factor is expected to increase by about 0.092 ps nm^{-2} km^{-1} or about 32 760 ps nm^{-1} at 1560 nm and 26 400 ps nm^{-1} at 1530 nm [19].

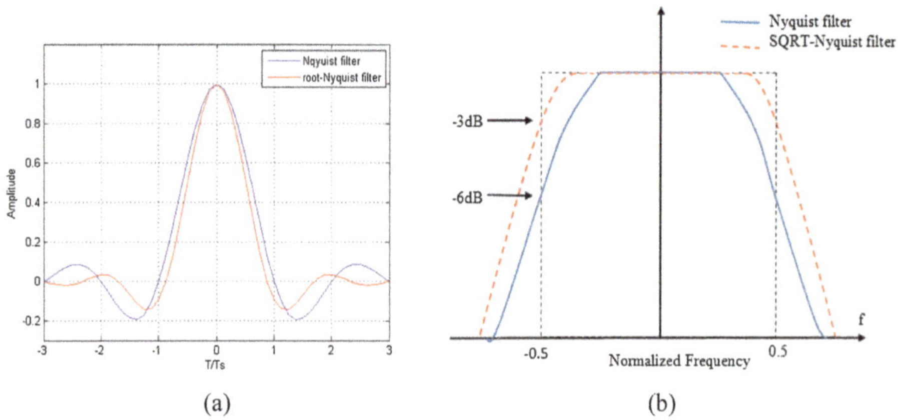

Figure 3.35. (a) Impulse and (b) corresponding frequency response of sinc Nyquist pulse shape or RRC Nyquist filters.

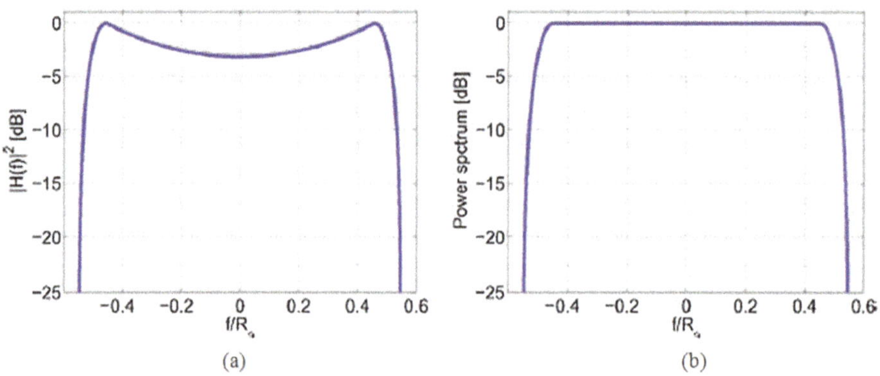

Figure 3.36. (a) Desired Nyquist filter for spectral equalization. (b) Output spectrum of the Nyquist filtered QPSK signal.

PMD tolerance. The worst case of deployed fiber for 2000 km would have a DGD of 75 ps or about three symbol periods for 25 GSy s^{-1} per sub-channel. So the PMD (mean all-order DGD) tolerance is 25 ps.

SOP rotation speed. According to the 100 Gbps experiments, SOP rotation can be up to 10 KHz, so we take the same spec as the 100 G system (figure 3.36).

Modulation format. PDM-QPSK for long distance transmission and PDM-16QAM for metro applications.

Spectral efficiency. Compared to the 100 G system, this increases by a factor of 2. Both Nyquist WDM and Co-OFDM can fulfill this. However, this depends on the technological and economical requirements that would determine the suitability of the technology for optical network deployment.

Table 3.2 tabulates the system specifications of various transmission structures with the parameters of sub-systems, in particular when the comb generators

Table 3.2. Specifications of the 1 Tbps offline system.

Parameter Technique	Superchannel RCFS comb gen.	Superchannel Nonlinear comb gen.	Some specs	Remarks
Bit rate	1, 2, ... N Tbps (the whole C-band)	1.0 Tbps, 2.0... N Tbps	~1.28 Tbps @ 28–32 GBd	20% OH for OTN, FEC
Number of ECLs	1	$N \times 2$		
Nyquist roll-off	0.1 or less	0.1 or less		DAC pre-equalization required
Baud rate (GBd)	28–32	28–32	28, 30 or 31.5 GBd	Pending on FEC coding allowance
Transmission distance	2500	2500	1200 (16 span) ~ 2000 km (25 spans) 2500 km (30 spans)	20% FEC req. for long distance application
			500 km	Metro application
Modulation format	QPSK/16QAM	QPSK/16QAM	Multi-carrier Nyquist WDM PDM-DQPSK/QAM	For long distance
				For long distance
			Multi-carrier Nyquist WDM PDM-16QAM	For metro
Channel spacing			4×50 GHz	For long distance
			2×50 GHz	For long distance
Launch power	$\ll$0 dBm if 20 Tbps is used		~ −3 to 1 dBm lower if $N > 2$	Depending on QPSK/16QAM and long distance/metro can be different
B2B ROSNR @ 2e-2 (BOL) (dB)	14.5	14.5	15 dB for DQPSK 22 dB for 16QAM	1 dB hardware penalty 1 dB narrow filtering penalty
Fiber type	SSMF G.652 (or G.655)	SSMF G.652 (or 655)	G.652 SSMF	
Span loss	22	22	22 dB (80 km)	

(Continued)

Table 3.2. (*Continued*)

Parameter Technique	Superchannel RCFS comb gen.	Superchannel Nonlinear comb gen.	Some specs	Remarks
Amplifier	EDFA (G > 22 dB); NF < 5 dB		EDFA (OAU or OBU)	
BER	2e-3	2e-3	Pre-FEC 2e-2 (20%) or 1e-3 classic FEC (7%)	
CD penalty (dB)			0 dB @ ±3000 ps nm^{-1} <0.3 dB @ ±30 000 ps nm^{-1}	16.8 ps nm^{-1} km^{-1} and 0.092 ps nm^{-2} km^{-1}
PMD penalty (DGD)			0.5 dB @ 75 ps, 2.5 symbol periods	
SOP rotation speed	10 kHz	10 kHz	10 kHz	OPLL may require due to oscillation of the LO carrier
Filters cascaded penalty			<1 dB @ 12 pcs WSS	
Driver linearity	required	required	THD < 3%	16QAM even more strict

employed use either re-circulating or nonlinear generation techniques. The DSP reception and offline digital signal processing is integrated in these systems.

3.6.4 System structure: DSP-based coherent receiver

A possible structure of a superchannel transmission system can be depicted in figure 3.37. At the transmitter the data inputs can be inserted into the pulse shaping and individual data streams can be formed. A DAC can be used to shape the pulse with Nyquist equivalent shape, the raised cosine form whose spectra also follow a raised cosine with the roll-off factor β varying from 0.1 to 0.5. If this off factor takes the value of 0.1, the spectra would follow an approximately rectangular shape. A comb generator can be used to generate equally spaced sub-carriers for the superchannel from a single carrier laser source, commonly a very narrow band external cavity laser of linewidth of about 100 KHz. These comb generated sub-carriers (see figure 3.39) are then demultiplexed into sub-carriers and fed into a bank of I–Q optical modulators, and Nyquist pulse shaped pulse sequences as the output of the DAC are then employed to modulate these sub-carriers to form the super-channels at the output of an optical multiplexer, shown in the block on the left side of figure 3.37. More details of the transmitter for superchannels are shown in figure 3.38. It is noted that the generation of a comb source can be re-circulating of the shifting of the original carrier around a closed loop. The frequency shift is the spacing frequency between the sub-carriers. So the *N*th sub-carrier would be the *N*th time circulation of the original carrier. There would be superimposition of noise due to the ASE incorporated in the loop, thus this would be minimized by inserting an optical filter into the loop whose bandwidth would be the same as or wider than that of the superchannel.

The fiber transmission line is an optically amplified optical fiber multi-span without incorporating any dispersion compensating fibers. Thus the transmission is highly dispersive. The broadening of a 40 ps width pulse would spread across at least an 80 to 100 symbol period after propagating over 3000 km of SSMF. Thus one can assume that the pulse launched into the fiber of the first span would be considered to be an impulse compared to that after 3000 km SSMF propagation.

After the propagation over the multi-span non-DCF line, the transmitted sub-channels are demuxed via a wavelength splitter into individual sub-channels, with

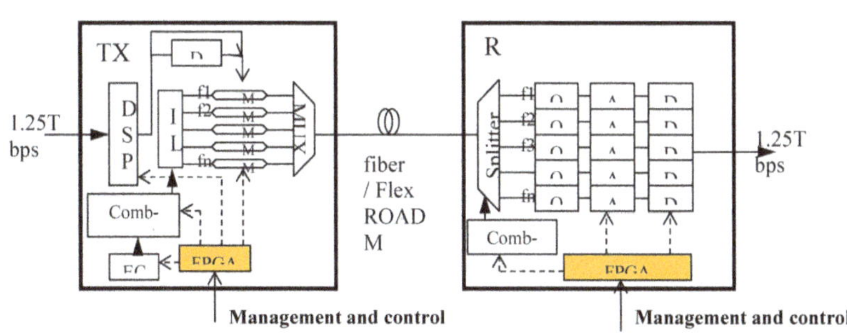

Figure 3.37. The structure of a superchannel Tbps transmission system.

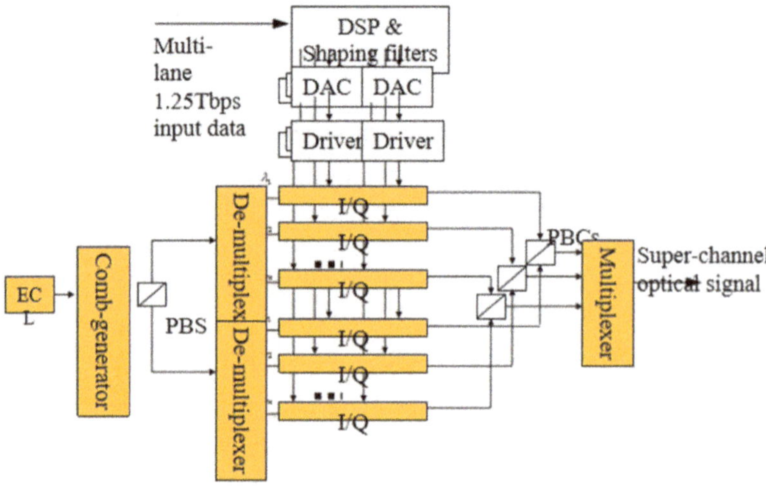

Figure 3.38. Generic detailed architecture of a superchannel transmitter. PBC = polarization beam combiner; PBS = polarization beam splitter; I/Q = in-phase/quadrature phase; DAC = digital to analog convertor; ECL = external cavity laser.

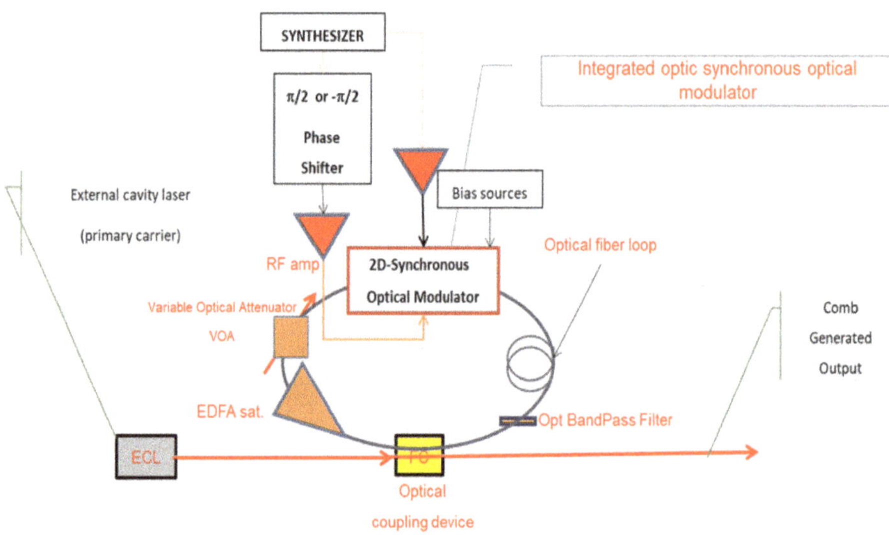

Figure 3.39. (a) Block diagram of a re-circulating frequency shifting comb generator and (b) a typical generated spectrum of the comb generator with 28 GHz spacing between channels over more than 5 nm in the spectral region and an ~30 dB carrier-to-noise ratio (CNR).

minimum crosstalk. Each sub-channel is then coherently mixed with an LO which is generated from another comb source incorporating an OPLL to lock the comb into that of the sub-carriers of the superchannel. Thus a comb generator is indicated on the right side, the reception system, of figure 3.37. The coherently mixed sub-channels are then detected by a balanced receiver, and then electronically amplified

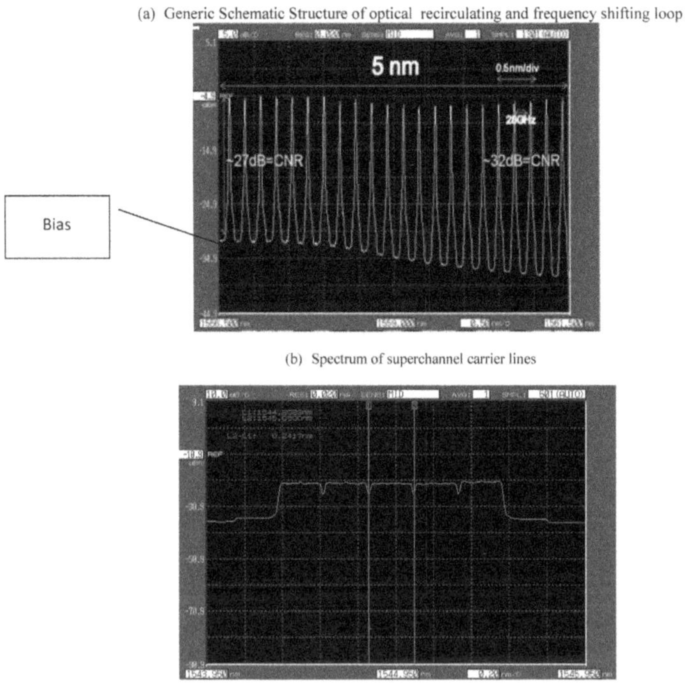

Figure 3.40. Selected five sub-carriers with modulation.

and fed into the sampler and ADC. The digital signals are then processed in the DSP of each sub-channel system or parallel and interconnected DSP system (figure 3.39). In these DSPs the sequence of processing algorithms is employed to recover the carrier phase and hence the clock recovery, compensating for the linear and nonlinear dispersion, and the evaluation of the BER versus different parameters such as OSNR, etc. Figure 3.40 shows the modulated spectra of five channels whose sub-carriers are selected from the multiple sub-carrier source of figure 3.39. The modulation is QPSK with Nyquist pulse shaping.

3.7 Timing recovery in the Nyquist QAM channel

Nyquist pulse shaping is one of the efficient methods to pack adjacent sub-channels into a superchannel. The timing recovery of such a Nyquist sub-channel is critical for sampling the data received and improving the transmission performance. Timing recovery can be performed either before or after the PMD compensator. The phase detector scheme is shown in figure 3.41, and is the Godard type [20], which is a first-order linear scheme. After CD compensation (CD^{-1} blocks), the signal is sent to a state-of-polarization (SOP) modifier to improve the clock extraction. The clock performance of the NRZ QPSK signal in the presence of a first-order PMD is characterized by a differential group delay (DGD) and an azimuth is present. An azimuth of 45° and a DGD of a half symbol/unit interval (UI) completely destroys the clock tone. Therefore, the SOP modifier is required for enabling the clock

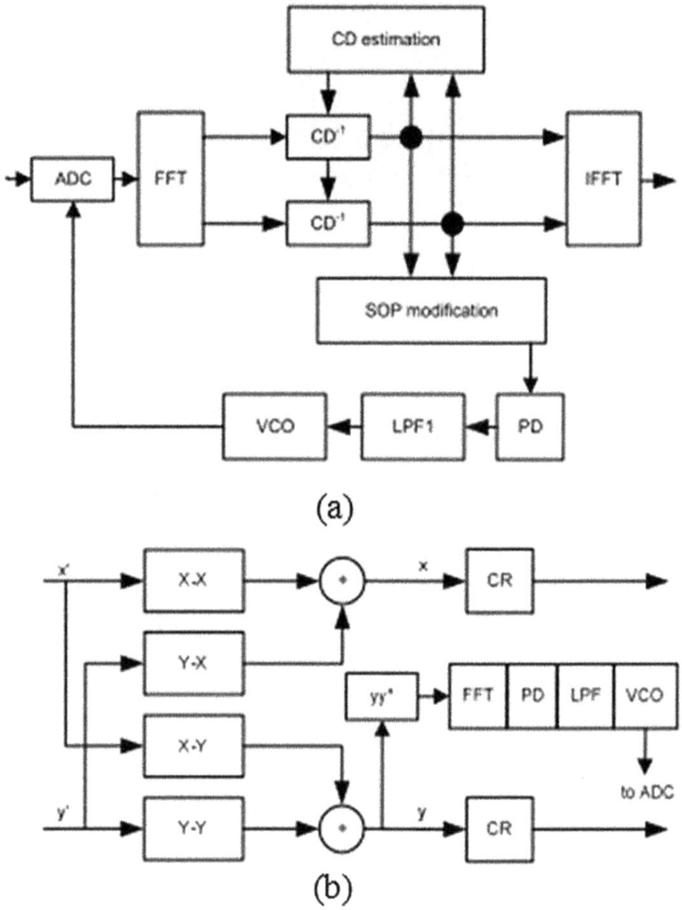

Figure 3.41. (a) Godard phase detector algorithm. (b) Forth-order PPD. VCO = voltage control oscillator; FFT = fast Fourier transform; IFFT = inverse FFT; CD = chromatic dispersion; SOP = state of polarization.

extraction. In practical systems, a raised cosine filter is used to generate Nyquist pulses. A filter pulse response is defined by two parameters, the roll-off factor (ROF) β and the symbol period T_s and is described by taking the inverse Fourier transform. The Godard phase detector cannot recover the carrier phase even with small β, thus the channel spectra are close to rectangular. A higher order phase detector must be used to effectively recover the timing clock period, as shown in [21]. A fourth power law PD (4PPD) with pre-filtering, presented in [22] and as shown in figure 3.41(b), can deal with small β values.

The 4PPD operates by first splitting and forming the combination of the components of the *X*- and *Y*-polarized channels, then conducting the frequency domain detection and regenerating, through the VCO, the frequency shift required for the ADC to ensure the sampling timing is correct with the received sample for processing.

The use of such a phase detector in coherent optical receivers requires a large hardware contribution. Due to PMD effects the direct implementation of this method before PMD compensation is almost impossible. Therefore, the 4PPD implementation in the frequency domain after the PMD compensation is proven to be the most effective and performs well even in the most extreme cases with ROF equal to zero.

3.8 128 Gbps 16QAM superchannel transmission

This is an experimental set-up by Dong *et al* [23] for the generation and transmission of six channels carrying 128 Gbit s^{-1} under the modulation format and a polarization multiplexing PDM-16QAM signal. The two 16 Gbd electrical 16QAM signals are generated from the two arbitrary waveform generators. Laser sources at 1550.10 nm and the second source with a frequency spacing of 0.384 nm (48 GHz) is generated from two external cavity lasers, each with a linewidth less than 100 kHz and an output power of 14.5 dBm, respectively. Two I/Q MODs are used to modulate the two optical carriers with I and Q components of the 64 Gb s^{-1} (16 Gbd) electrical 16QAM signals after the power amplification using four broadband electrical amplifiers/drivers, respectively. Two phase shifters with the bandwidth of 5k–22.5 GHz provide a two-symbol extra delay to de-correlate the identical patterns. For the operation to generate 16QAM, the two parallel Mach–Zehnder modulators (MZMs) for I/Q modulation are both biased at the null point and driven at full swing to achieve zero-chirp and phase modulation. The phase difference between the upper and the lower branch of I/Q MZIM is also controlled at the null point. The data input is shaped so that about a 0.99 roll-off factor of raised cosine pulse shape could be generated.

The power of the signal is boosted using polarization-maintaining EDFAs. The transmitted optical channels are then mixed with a local oscillator and polarization demultiplexed via a $\pi/2$ hybrid coupler, then the I and Q components are detected by four pairs of balanced photodetectors (PDP). They are then trans-impedance amplified and re-sampled.

The sampled data are then processed in the following sequence: CD compensation, clock recovery, re-sampling and going through classical CMA, a three-stage CMA, frequency offset compensation, feed forward phase equalization, LMS equalizer (LMSE) and then differentially detected to avoid cycle slip effects.

Furthermore, the detailed processing [23] for the electrical polarization recovery is achieved using a three-stage blind equalization scheme. (i) First, the clock is extracted using the 'square and filter' method, and then the digital signal is re-sampled at twice the baud rate based on the recovery clock. (ii) Second, a $T/2$-spaced time domain finite impulse response (FIR) filter is used for the compensation of CD, where the filter coefficients are calculated from the known fiber CD transfer function using the frequency domain truncation method. (iii) Third, adaptive filters employing two complex-valued, 13 tap coefficients and partial $T/2$-space are employed to retrieve the modulus of the 16QAM signal.

The two adaptive FIR filters are based on the classic constant modulus algorithm (CMA) and followed by a three-stage CMA, to realize multi-modulus recovery and polarization de-multiplexing. The carrier recovery is performed in the subsequent

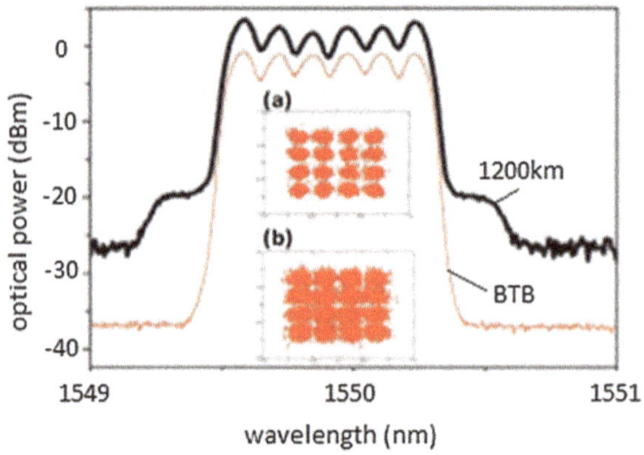

Figure 3.42. Spectra of Nyquist channels, 6 × 128G PDM-16QAM back-to-back and 1200 km transmission.

step, where the fourth power is used to estimate the frequency offset between the LO and the received optical signal. The phase recovery is obtained by feed forward and the least mean squares (LMS) algorithms for local oscillator frequency offset (LOFO) compensation. Finally, differential decoding is used for BER calculation after decision.

The spectra of the six Nyquist channels under B2B and 1200 km SSMF transmission are shown in figure 3.42. It is noted that even for the 0.01 roll-off of Nyquist channels, we do not observe the flatness of individual channels in the diagram. The constellation as expected would require a significant amount of equalization and processing.

On average, the achieved BER of 2e-3 with a launched power of −1.0 dBm over 1200 km SSMF non-dispersion compensation link was demonstrated by the authors of [16]. The link is determined by a re-circulating loop consisting of four spans with each span having a length of 80 km SSMF and an in-line EDFA. A wavelength selective switch (WSS) is employed wherever necessary to equalize the average power of the sub-channels. The optimum launched power is about −1 dBm for the six-sub-channel superchannel transmission.

3.9 450 Gb s^{-1} 32QAM Nyquist transmission systems

Further spectral packing of sub-channels in a superchannel can be done with Nyquist pulse shaping and pre-distortion or pre-equalization at the transmitting side. Zhou *et al* [24] recently demonstrated the generation and transmission of 450 Gb s^{-1} wavelength-division multiplexed (WDM) channels over the standard 50 GHz ITU-T grid optical network at a net spectral efficiency of 8.4 b s^{-1} Hz^{-1}. This result is accomplished by the use of Nyquist shaped, polarization division multiplexed (PDM) 32QAM, or 5 bits/symbol × 2 (polarized modes) × 45 GBd s^{-1} to give 450 Gb s^{-1}. Both pre- and post-transmission digital equalization techniques are employed to overcome the limitation of the DAC bandwidth. Nearly ideal Nyquist pulse shaping with a roll-off factor of 0.01 allows guard bands of only 200 MHz between sub-carriers. To mitigate the narrow optical filtering effects from the 50 GHz grid reconfigurable optical add–drop multiplexer (ROADM), a broadband optical pulse shaping method is employed. By combining electrical and optical shaping techniques, the transmission of 5×450 Gb s^{-1} PDM-Nyquist 32QAM on the 50 GHz grid over 800 km and one 50 GHz grid ROADM was proven with soft DSP equalization and processing. The symbol rate is set at 28 GSy s^{-1}.

It is noted that the transmission SSMF length is limited to 800 km due to the reduced Euclidean geometrical distance between the constellation points of 32QAM and by avoiding the accumulated ASE noise contributed by EDFA in each span. Raman optical amplifiers with distributed gain are used in a re-circulating loop of 100 km ultra-large area fibers. The BER performance for all five sub-channels is shown in figure 3.43(a) with an insert of the spectra of all sub-channels. Note the near flat spectrum of each sub-channel that indicates the near Nyquist pulse shaping. Figures 3.43(b) and (c) show the spectra of a single sub-channel before and after WSS with and without optical filtering that performs as spectral shaping. The original pulse shape can be compared with that displayed in figure 3.43. Further to this published work, Zhou *et al* have also time multiplexed 64QAM and transmitting over 1200 km [25], that is three circulating around the ring of 400 km of a 100 km span incorporating Raman pumped amplification. An approximately 5 dB penalty between 32QAM and 64QAM in receiver sensitivity at the same FEC BER of 6e-3 is obtained. Note that electrical time division multiplexing of digital sequences from the arbitrary waveform generators was implemented by interleaving such that a higher symbol rate can be achieved. Furthermore, 20% soft decision FEC using a quasi-cyclic LDPC code is employed to achieve a BER threshold of 2.4e-2, thus the 20% extra-overhead is thus required on the symbol rate.

Simulated results by Bosco *et al* [26] without using soft FEC also shows that the variation of the maximum reach distance with a BER of 2e-3 for capacity in the C-band (bottom axis) and spectral efficiency for PM-BPSK, PM-QPSK, PM-8QAM and PM-16QAM, as shown in figure 3.44 for SSMF non-DCF optical transmission lines as well as non-zero dispersion shifted fibers (NZ-DSF) indicated by dashed lines.

3.10 DSP-based heterodyne coherent reception systems

We have so far described optical transmission systems under homodyne coherent reception, that is when the frequencies of the lightwave carriers and that of the local oscillator are equal. The original motivation for using homodyne detection is to eliminate the 3 dB degradation compared to the heterodyne technique, and this is possible under the DSP-based reception algorithm to avoid the difficulties of locking of the local oscillator and the channel carrier. Under the classical heterodyne reception the 'at least' 3 dB loss comes from the splitting of the received signals in the electrical domain and then multiplied by a sinusoidal cosine and sine RF oscillator to extract the in-phase and quadrature (I–Q or I/Q) components.

So far we have discussed homodyne reception DSP-based optical transmission systems, which are considered for extensive deployment in commercial coherent communication systems for 100 G, 400 G, 1 T or beyond. However, with the

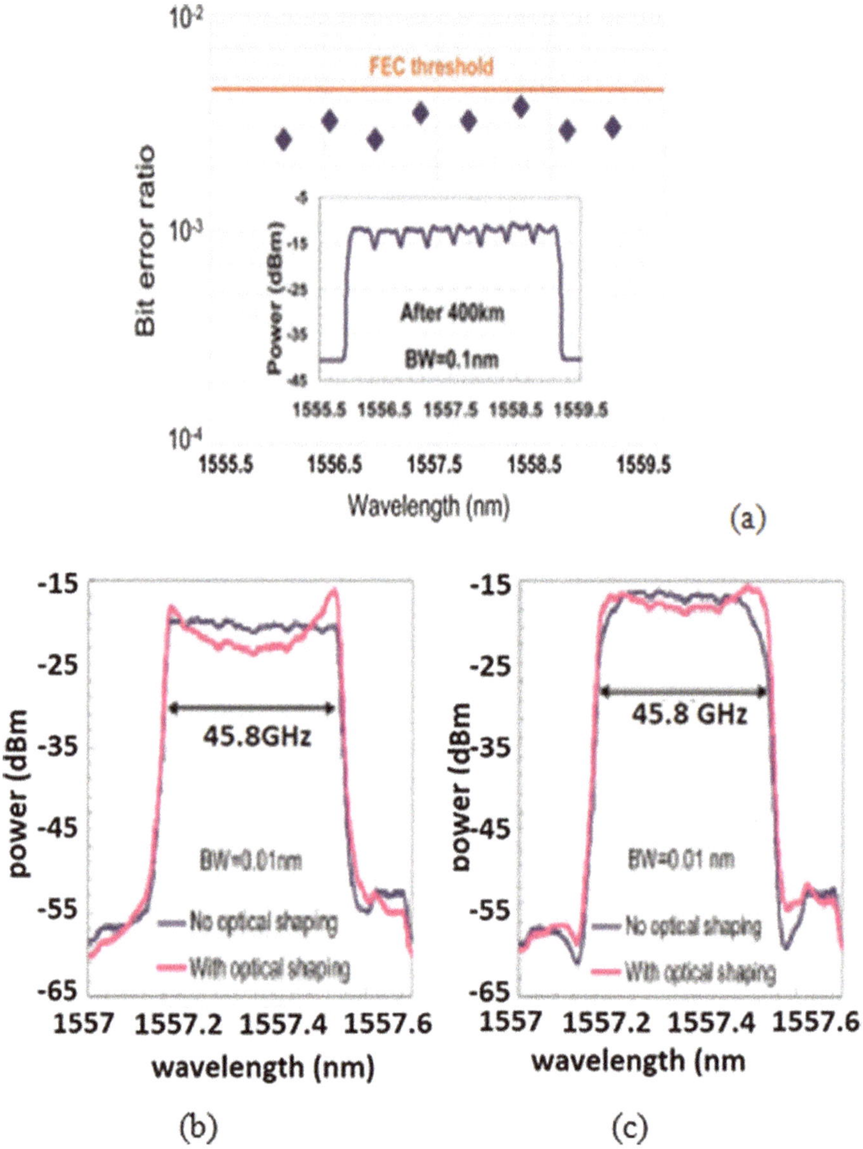

Figure 3.43. (a) Spectra of five sub-channels of a 450 Gb s^{-1} channel, that is 5 × 450 Gb s^{-1} super-channels after 400 km (one loop circulating) of 5 × 80 km plus Raman amplification. BER versus wavelength and FEC threshold at 2.3e-3. Spectrum of a single sub-channel (b) before and (c) after WSS with and without optical shaping by optical filters (reproduced from [17] with permission). Note: blue line for original spectra and pink line for spectra after spectral shaping.

development of large-bandwidth and high speed electronic ADCs and photodetectors (PDs), once again, coherent detection with digital signal processing (DSP) has allowed the mitigation of impairments in optical transmission which can be compensated by equalization in the electrical domain. For homodyne detection in polarization division multiplexing (PDM) systems, the I/Q components

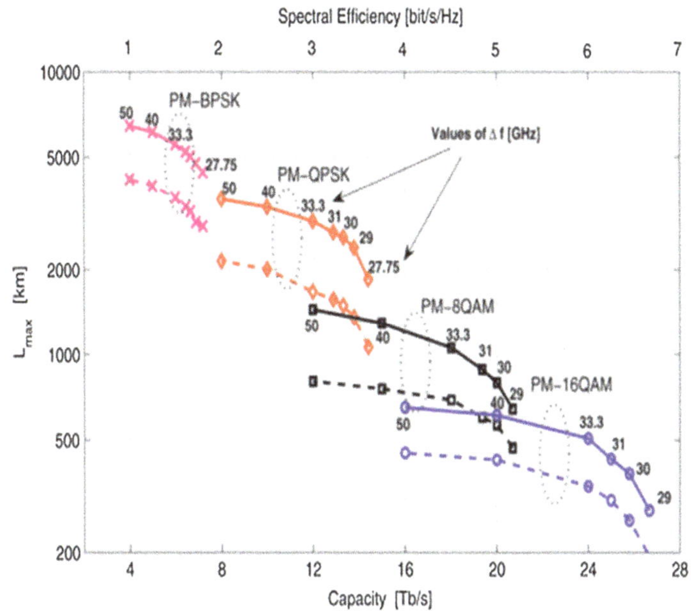

Figure 3.44. Transmission reach distance variation with respect to spectral efficiency and total capacity of superchannels. Different QAM schemes are indicated. PM = polarization multiplexing. The numbers are for bandwidth of the sub-channel (reproduced from [18]).

of each polarization state should be separated in the optical domain with full information. Thus, four balanced PD pairs incorporated with a photonic dual-hybrid structure and four-channel time-delay synchronized ADCs are required.

By up-converting I and Q components to the intermediate frequency (IF) at the same time, not only can heterodyne coherent detection halve the number of balanced PDs and ADCs of the coherent receiver, but there is also no need to consider the delays between the I and Q components in the PDM signal. Therefore, the four output ports of the optical hybrid can also be halved accordingly. However, this heterodyne technique can possibly be restricted by the bandwidth of the photo-detectors (PDs). Furthermore, the external down conversion of the IF signals may enhance the complexity of the reception system. Currently, the tremendous progress in increasing the sampling rate and bandwidth of ADCs and PDs gives a high possibility to exploit a simplified heterodyne detection. With large-bandwidth PDs and ADCs, down conversion of the IF, I/Q separation of quadrature signal, and equalization for the PDM and nonlinear effect can all be realized in the digital domain, the DSP following the ADCs. A heterodyne detection in the transmission system is a limited 5 Gb s^{-1} 4QAM signal over 20 km in [27] and a limited 20 Mbd 64QAM and 128QAM over 525 km in [28], and then reaching Tbps by Dong *et al* [29]. High-order modulation formats, such as PDM-16QAM and PMD-64QAM, taking advantage of high capacity, can offer spectral effectiveness for 100 G or beyond by adopting heterodyne detection.

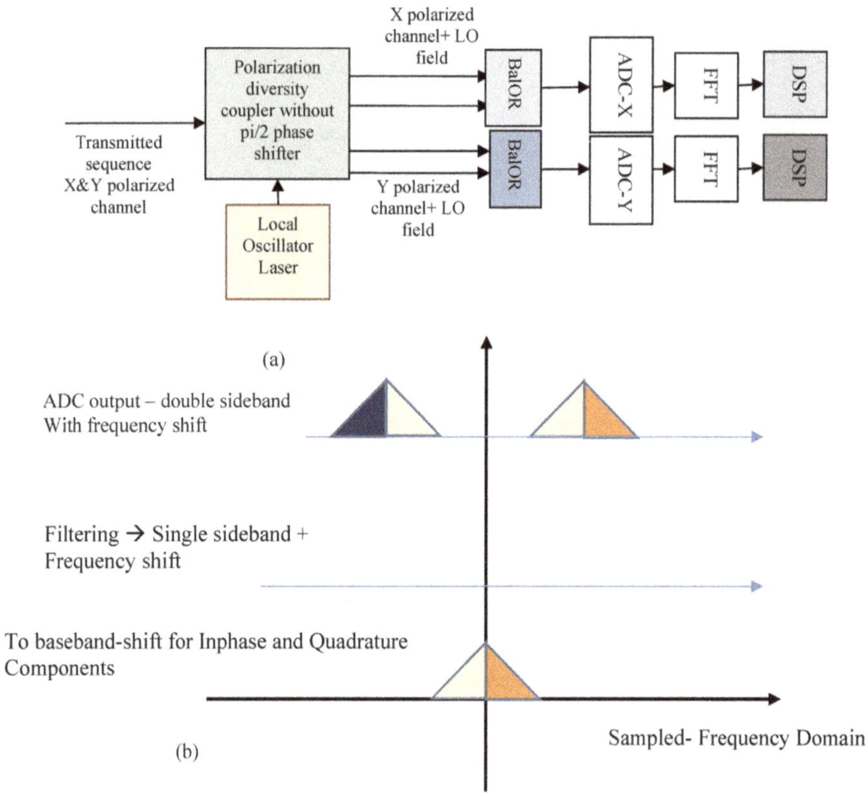

Figure 3.45. Principles of heterodyne coherent reception: (a) principal blocks and (b) spectra to recover baseband signals.

For the 100 G or beyond coherent system with a required transmission distance shorter than 1000 km, the inferiority of the SNR sensitivity in heterodyne detection is not so obvious. Conversely, a smaller number of ADCs and easy implementation of the DSP for IF down conversion make heterodyne detection a potential candidate for 100 G or beyond transmission system. Figure 3.45(a) shows the schematic diagram of the heterodyne reception with digital processing. The quadrature amplitude modulated signals are transmitted and imposed onto the $\pi/2$ hybrid coupler which is now simplified and there is no $\pi/2$ phase shifter compared to the hybrid coupler. The coherent mixing of the LO and the modulated channel would result in time domain signals which are an RF envelope covering the lightwaves and limited within the symbol period. Thus both the real and imaginary parts appear in the electronic signals produced after the balanced detection in the PDPs in which the beating happens (see figure 3.45(b)). The electronic currents produced after the PDPs are then amplified via the TIA to produce voltage-level signals which are conditioned to appropriate levels so that they can be sampled by the ADCs. Hence performing the FFT will produce a two-sided spectrum which exhibits frequency shifting of the baseband to the RF or IF frequency. Both the in-phase and quadrature components are embedded in the two-sided spectrum. One sideband of the spectrum can be used to extract the I and Q parts of the QAM signals for further processing. Ideally, the compensation of the CD should be performed in the first stage and then be followed by carrier phase recovery and then re-sampling with correct timing. Section 3.17 depicts the schematic of a balanced optical receiver and its principles of operation.

The following sequence of processing in the digital domain was conducted: (i) sampling in ADC and conversion from an analog voltage level to digital sampled states; (ii) performing FFT to obtain the frequency domain spectrum; (iii) extracting a one-sided spectrum and performing a frequency shifting to obtain the baseband samples of the spectrum; (iv) re-sampling with two times the sampling rate; (v) compensation of CD; (vi) carrier phase and clock recovery; (vii) using the recovered clock to resample the data sequence; (viii) conducting normal CMA and three-stage CMA to obtain the initial constellation; (ix) frequency equalization of the frequency difference of the LO; (x) feed forward phase equalization, and LMS equalization for PMD compensation; then finally (xi) differential decoding of the samples and symbol to determine the transmission performance BER with respect to certain OSNR measured in the optical domain and launched power at the input of the transmission line.

It is further noted that for the equalization based on DSP, a *T*/2-spaced time domain finite impulse response (FIR) filter is used for the compensation of chromatic dispersion (CD). The two complex-valued, 13-tap, *T*/2-spaced adaptive FIR filters are based on the classic constant modulus algorithm (CMA) followed by three-stage CMA, to realize multi-modulus recovery and polarization de-multiplexing.

The back-to-back BER versus the OSNR of the heterodyne reception optical transmission system is shown in figure 3.46 for PDM-16QAM 128 Gb s^{-1} per sub-channel with different IF and compared to the homodyne detection scheme. It is noted that there is a 3 dB penalty of heterodyne compared to homodyne reception due to the fact that only a one-sided spectrum is employed. Thus at a BER of 4e-3 the required OSNR is about 23 dB. The transmission distance is 720 km of SSMF incorporating EDFA and non-DCF.

3.11 Remarks

The total information capacity transmitted over an SSMF can be increased by spectral packing of sub-channels by pulse shaping techniques and DSP processing algorithms as well soft FEC and/or pre-emphasis/distortion of the transmitting signals so as to allow a higher order of modulation QAM to be realized. Thus digital processing in real-time and digitally based coherent optical transmission systems have been proven to be the most modern transmission systems for the global Internet in the near future. The principal challenges are now lying on the realization of application specific integrated circuits (ASIC) using microelectronics technology or ultra-fast field programmable gate array (FPGA) based systems. The concepts of processing in the digital domain have been proven in offline systems and thus more

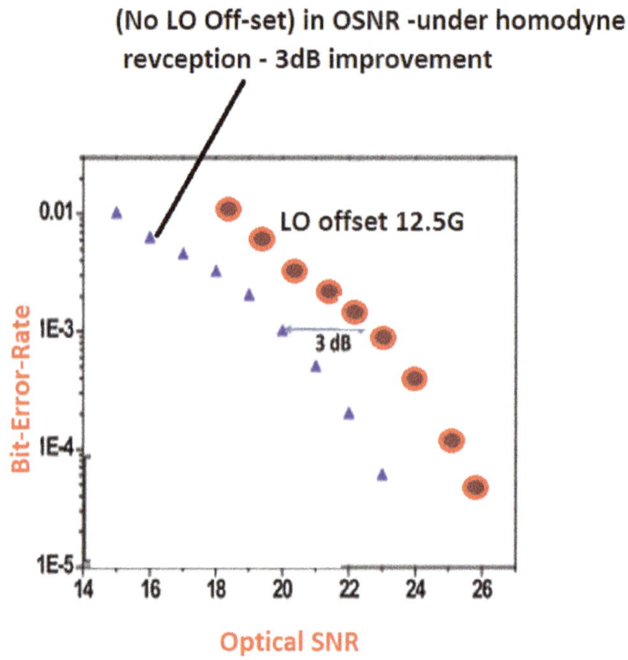

Figure 3.46. Back-to-back BER versus OSNR of the heterodyne reception system of a PDM-16QAM 8 × 112 Gbps superchannel with different IF frequency (reproduced from [2] with permission).

efficient algorithms are required for real-time systems. These remain current research topics and engineering issues. Higher symbol rates may be possible when wider bandwidth optical modulators and DAC and ADC systems are available, e.g. grapheme plasmonic silicon modulators [30] and their integration with micro-electronic DAC and ADC DSP systems.

The processing algorithms for QAM vary from level to level but essentially they can be based on the number of circles existing in multi-level QAM compared to a mono-cycle constellation of the QPSK scheme. The algorithms developed for QPSK can be extended and modified for higher level circular constellations.

3.12 PAM4 IM/DD systems

3.12.1 Generating PAM4 signals

In generating PAM4 signals the most common method is to use the arbitrary generator (AGW) or DSP generated outputs and then feed them through a DAC so that the analog signal waveforms can be amplified in an RF amplifier and then be fed to an optical modulator. Complementary channels can also be available for differential drives. The main problem in this technique is the amount of power consumption, in particular when the sampling rate is increased as a cubic function of the sampling.

In order to avoid such power consumption, a photonic technique can be employed with two optical modulation paths. One modulator is driven with NRZ and the other also with NRZ but with 6 dB amplitude reduction. Thus the PAM4 format can be generated as recently reported. Another method is to use a multi-section electrode of an MZIM. Each section is driven by a phase element of a phase pattern which is equivalent to the digital waveform. This section describes these advanced generation techniques of the PAM4 sequence so as to double the bit rate of the transmission systems.

3.12.2 DSP-based PAM4 generation for low cost broadband channels

Two main contemporary approaches have been reported recently [31, 45], one approach in the optical domain and the other in the analog electronic domain. In the optical domain approach two parallel optical paths, incorporating an EA modulation path, whose amplitudes are 3 dB different in intensity are used to generate the PAM4 signal channels. On the other hand, the electronic approach is achieved by generating a multi-section electrode which is placed on one optical path followed by phase modulation of the sections of the electrode by electronic amplified signals whose amplitudes are digitally equivalent to DAC generated bits.

3.12.2.1 DAC-free PAM4 generation by dual optical path modulation

A DAC-free optical transmitter is preferred in low cost transmission links as the power consumption can be maximally reduced. One way is to use two optical modulators which are driven by a pair of synchronized signals of the same RF broad bandwidth. Their amplitudes are 3 dB different in power, as shown in figure 3.47(a).

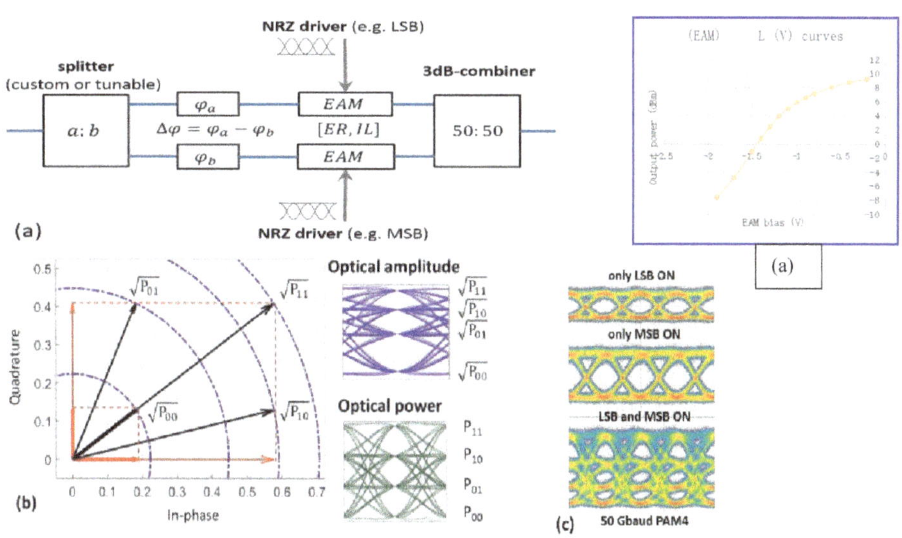

Figure 3.47. (a) Block diagram of the DAC-less Tx. (b) Example of equidistant PAM4 generation using the first quadrant of the complex plane where the power split ratio *a*:*b* was chosen as 0.33:0.66, $\varphi = 90°$, and the EAMs exhibit an absorption factor which is limited to 10 dB or 10% in terms of power of the guided optical waves passing through the waveguide branch. In this special case, the optical power eye openings decrease with decreasing extinction ratio (ER) but they remain equidistant, without changing φ. (c) Optical eyes for the DAC-free transmitter at 50 GBd [45]. (d) Transfer curve of EAM modulator—output power versus applied voltage across the waveguide on the absorption region.

Two optical amplitude modulators of electro-absorption (EAM) type are fabricated by deposition of Ge thin-film on top of Si waveguides. Thus Ge can act as an absorber that is the same as a photodetector in the C-band window. Thus this acts as an in-line EAM. The phase tuning sections placed in front of the EAM can be added so that the tuning of the phase synchronization can be made. The outputs of the two EAMs are combined to give PAM4 signals. Therefore, QAM signals, for example 16QAM, can be generated if a phase shifter is placed in cascade with one EAM, i.e. $I/2$ phase shifting, hence complex signals can be generated. The phasor diagram and the eye diagrams can be generated as depicted in figure 3.47(b). Much more complex signal channels can be generated if DACs are used to alter the patterns feeding into the EAM. It is observed that the PAM4 eye diagram is amplitude-dependent because the delay time is amplitude-dependent on the electro-absorption effects in the EAM. Hence a tuning in the phase shifters placed in front of the EAM branches would balance the PAM4 pattern. More details of the operation of the DAC-free PAM4 TX can be found in [45]. Figure 3.47(d) depicts the power transfer as a function of the applied voltage to the electrode of the EAM.

3.12.2.2 Multi-sectional DAC-free modulator

Multi-section electrodes can be used to drive the optical path of an MZIM [31] as shown in figures 3.48(a) and (b). This structure is different than the DAC-less EAM PAM4 modulator shown in figure 3.47. The electrodes can be single-drive or deferentially driven in figure 3.48(b). The input impedance of these sectional electrodes must be impedance matched to the transmission line characteristic impedance. The main problem of this multi-sectional approach is the impedance matching, thus the delay can be different between the signal bits entering the electrodes so there would be distortion. Furthermore, the electronic delay would also be another additional difficulty. The analog electronics can also be more complicated compared to the DAC-free technique.

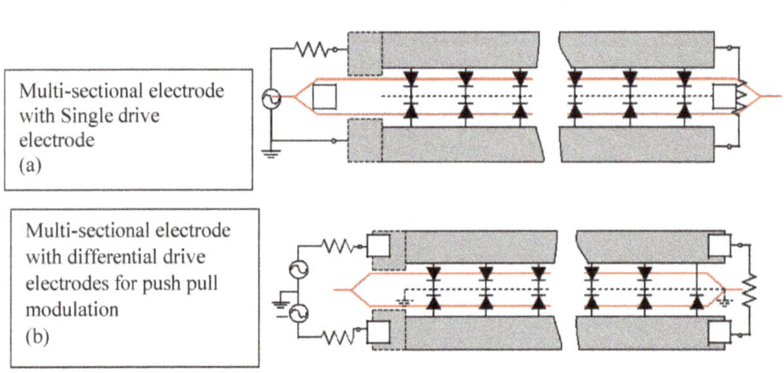

Figure 3.48. Schematic of (a) a series push–pull modulator which can be characterized by two ports and (b) a dual-drive modulator which requires four-port characterization. The dual-drive modulator shown here also has a p–n junction connected back-to-back and has a virtual ground at the middle due to its longitudinal symmetry. The red lines represent optical waveguides forming the MZMs.

3.12.3 PAM4 systems

This section gives recent development technologies for the ultra-high capacity transmission link in which modulation format PAM-4 (pulse amplitude modulation -4) of four levels in pulse amplitude and direct detection. These links offer transmission rates to higher than 100 Gbps using opto-electronic components whose 3 dB bandwidth is limited. As an example, opto-electronic components which are commonly employed in 10 Gbps transmission can be extended to 25 Gbps with processing by DSP. Thence higher modulation format PAM-4 can boost to even higher bit rate.

3.12.3.1 112 Gbps PAM4-OM4 by band-limited 10 G components

Low cost modules for data center networking (DCN) are critical for lowering its cost, particularly in distributed DCN. One way to lower the cost of the high speed module is to use components for 10 Gbps but for 100 G (25 G × 4 lanes) equalized by an MLSE algorithm implemented in the DSP processing. This sub-section presents the technique of this low cost approach. The system set-up is shown in figure 3.49. The 10 G components can be from manufacturer 1 (Sumitomo) or manufacturer 2 (EML = EA modulator integrated with DFB laser), also manufacturer 2's direct modulation laser (DML), which is a laser modulated by direct

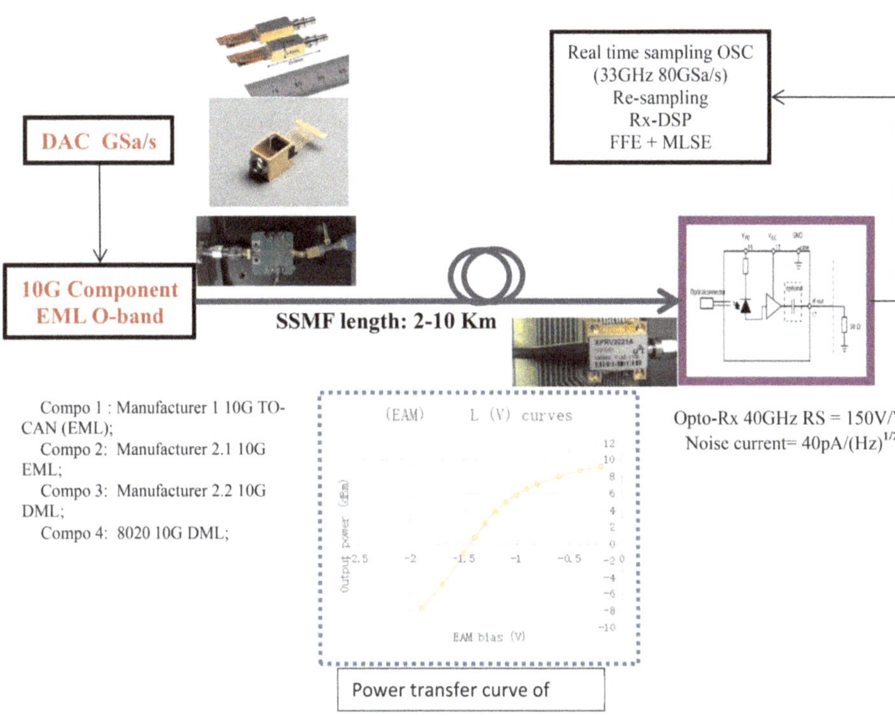

Figure 3.49. Low cost 100 G transmission system using 10 G components. Inset at bottom: EAM transfer curve-optical power output versus applied voltage and biasing of the EAM.

Table 3.3. Summary of the performance of different TOSA EML and DML components of different manufacturers (Smitomo and manufacturer 2 (a) and (b)). **B2B** = back-to-back; **ROP** = received optical power.

	Sumitomo EML	Manufacturer 2(a): EML	Manufacturer 2(a): DML	Manufacturer 2(b): DML
B2B ROP (dBm)	−2.7 @ 112 G	−7.6 @ 112 G	−1.8 @ 84 G	−2.8 @ 84 G
2 km ROP (dBm)	−2.5 @ 112 G	−7 @ 112 G	0.2 @ (84 G)	−3 @ (84 G)
5 km ROP (dBm)		−6.5 @ (112 G)		
10 km ROP (dBm)		−9.5 @ (84 G)		
B2B capacity (Gbps)	120	162	108	101
2 km capacity (GBps)	120	147	92	97
5 km capacity	—	130	—	—
10 km capacity	—	99	—	—

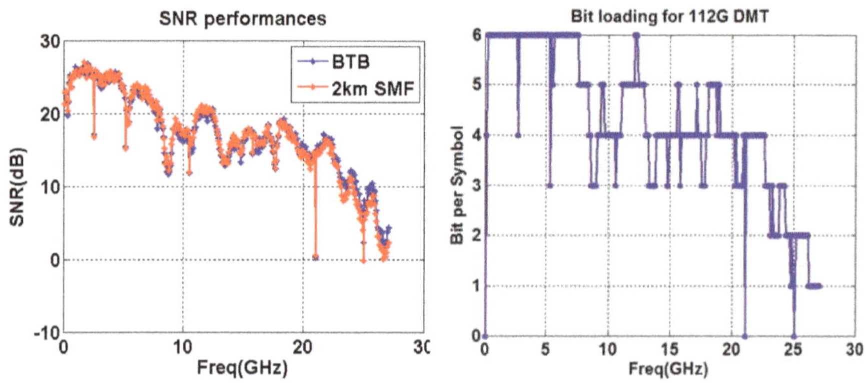

Power transfer curve of EAM.

Figure 3.50. Electrical SNR and the bit per symbol versus frequency to 30 GHz.

driving laser current, and another version of DML of manufacturer 2. These components are frequently used in 10 G transmission links and operating in the O-band (1310 nm spectral window). At the receiver side there is an optical receiver of bandwidth 40 GHz 3 dB passband (table 3.3). This receiving sub-system is composed of a high speed PD in tandem with a TIA with a spectral noise current density of 40 pA/sqrt(Hz) and a trans-impedance (TI) transfer power gain at mid-band of 150 V W^{-1}. The transfer power characteristics of the EAM are shown in the inset of figure 3.49.

As observed, the effective bandwidth can be 27 GHz with 10 dB and higher SNR. The number of bits per symbol that can be used varies from 2 to 6 over the 27 GHz band. Thus the order of higher modulation formats can be employed accordingly in the OFDM or DMT. Some impedance mismatch causes some degradation of the SNR observed in figure 3.50.

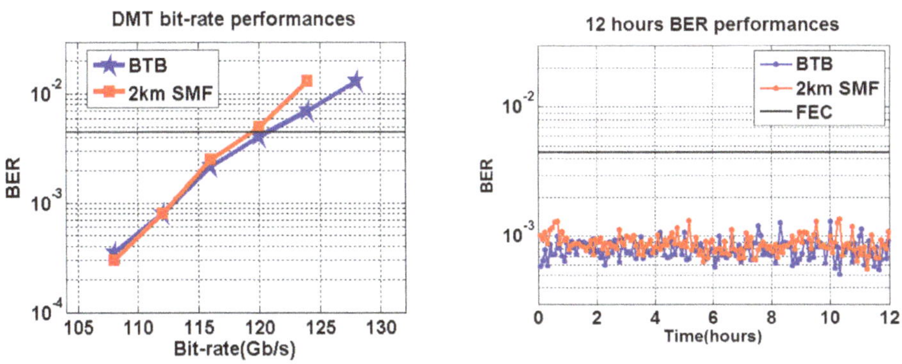

Notes: (i) The highest used frequency is around 27GHz; (ii) Power fading possibly caused by mismatching between the PCB and component?

Figure 3.51. Sumitomo TO-CAN TOSA: BER versus (a) receiver sensitivity (dBm) and (b) bit rate from 105 to 130 Gbps. B2B = back-to-back.

Table 3.4. Parameters of the DMT system at transmitter sub-system in terms of FFT size and DAC and ADC of different manufacturers.

AWG/DAC	Driver	TOSA/Mod.	ROSA	OSC/ADC
Angilent AWG (92 GS s^{-1}, 30 GHz)	SHF 807 (11 dB, 65 GHz)			Tek (200 GS S^{-1}, 70 GHz)
Angilent AWG (92 GS s^{-1}, 30 GHz)	SHF 827 (24 dB, 50 GHz)	Oclaro DDMZM (40 GHz)	U2t PD (70 GHz, no TIA)	Keysight (160 GS S^{-1}, 63 GHz)

The digitally modulated sequence is generated and output to drive the transmitter optical assembly (TOSA) via a DSP platform whose sampling rate is 100 GSa s^{-1}. 112 Gbps is successfully transmitted over 2 km SMF with a sensitivity of −2.5 dBm which is only a 0.2 dB power penalty after 2 km SM. Improvement is required to reduce the error-floor and the highest bit rate is 120 Gbps for both back-to-back and over 2 km lengths of SSMF. Figure 3.51 shows the ROP and capacity performance for the Sumitomo TO-CAN 10 G component transmitted over 2 km of SSMF. Physical components are shown in the insets of figure 3.49. For the TOSA of different manufacturers, their performance is listed in table 3.4.

3.12.3.2 PAM4 150 Gbps by band-limited components

It is understood that the bandwidth of electronic and integrated optic devices, in particular Si photonics, is limited to around 30–40 GHz (figure 3.52). So the total effective bit rate can be limited to around 80 Gbps for the NRZ modulation format (figure 3.53). The section, however, attempts to methodologically describe an NRZ, but with a PAM4 duobinary (PAM4 DuoB) using electrical filtering, not the optical

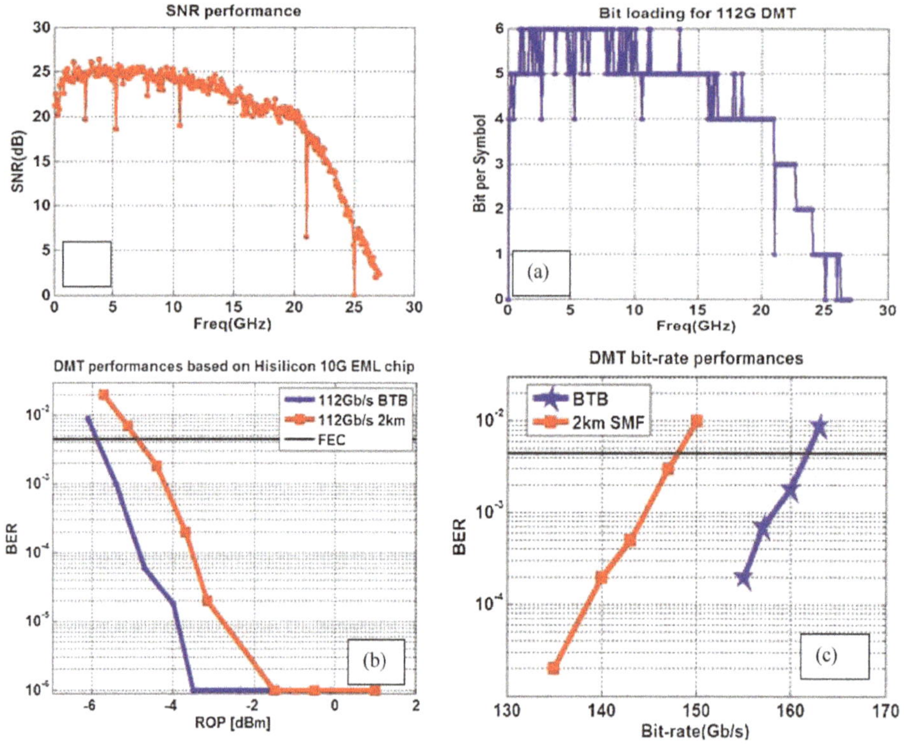

Figure 3.52. Manufacturer 2 EML: (a) SNR and (b) bit-loading of 10 G EML chip. The highest used frequency is around 27 GHz. Narrow power fading is caused by clock leakage. Manufacturer 2 EML: (c) ROP and (d) capacity performance for 2 km SMF.

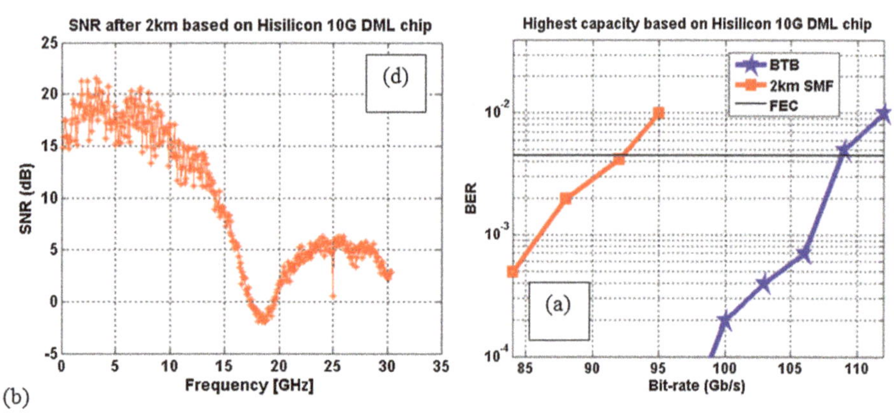

Figure 3.53. Manufacturer 2 10 GHz DML TOSA, performance for 2 km transmission: (a) frequency response of SNR and (b) BER versus bit rate—modulation DMT.

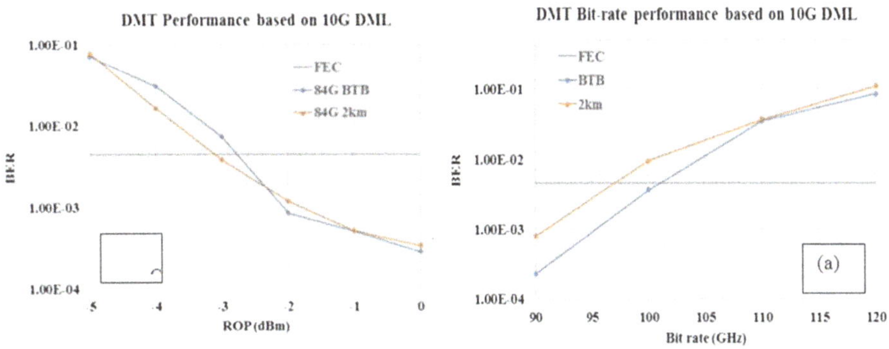

Figure 3.54. Manufacturer 2 DML: BER versus (a) receiver sensitivity (optical power input to the receiver) and (b) versus bit rate, therefore capacity performance for 2 km using the manufacturer 2 DML.

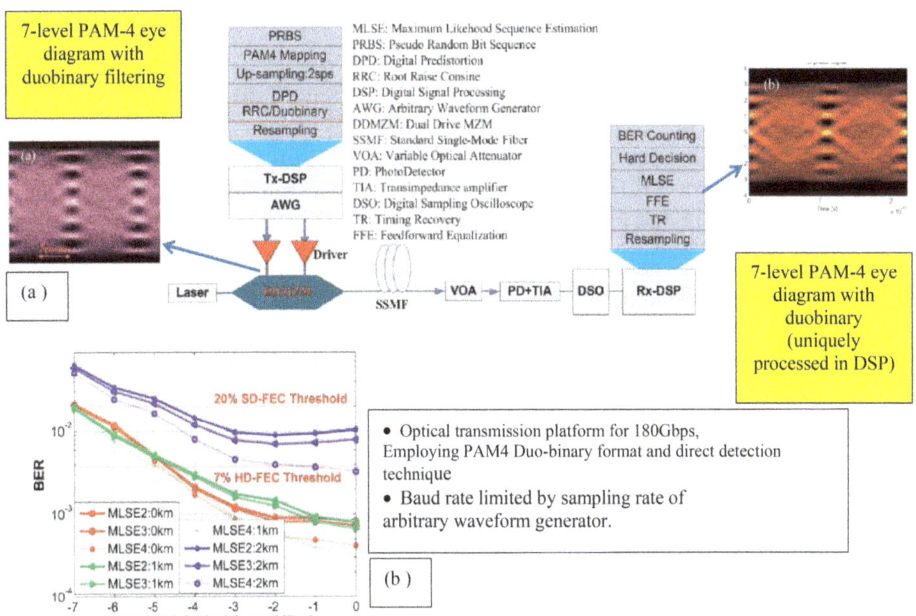

Figure 3.55. Demonstrated 180 Gb s^{-1} DB-PAM4. The bit rate is limited by the AWG (maximum 90 GS s^{-1}), MLSE to recover the pulse sequence.

duobinary phase [32], in association with DSP to demonstrate the bit rate to more than 150 Gbps (figure 3.54). The experimental transmission system is shown in figure 3.55(a). A laser is coupled to a dual electrode Mach–Zehnder interferometric modulator (MZIM) whose 3 dB bandwidth is about 32 GHz. The laser linewidth is about 2 MHz. The MZIM is driven by two radio frequency (RF) amplifiers fed by the analog signal output of the DAC of an arbitrary waveform generator (AWG). More details are given in the inset of figure 3.55(a). The transmission link is a 2 km SSMF. The transmitted signals are received by an optical receiver whose bandwidth is limited by the 3 dB bandwidth of 34 GHz of the trans-impedance amplifier (TIA). The signals are sampled and processed in the DSP with a sampling rate of 90 GSa s^{-1}.

The PAM4 duoB eye diagram shown in figure 3.55 indicates a seven-level eye diagram, and confirms the half-band filtering of the PAM4 eye. These eyes are at the transmitter input and at the front end of the DSP at the receiver. Both seem not to be affected very much after the transmission of the 2 km SSMF. The signal sequence is sampled and processed, recovered by maximum likelihood equalization (MLSE) and then BER is measured with respect to the variation of the receiver sensitivity by varying the attenuation coefficient of the variable optical attenuator (VOA). An FEC is used and the BER (1e-3 to 1e-4) is set at the level at which at 7% FEC and 20% soft FEC are allowed, as indicated in figure 3.55(b). The variations of the transmission distance from back-to-back to 2 km have a 1.0 km step. The sequence bit rate is 180 Gbps which is a baud rate of 90 GBd using the Nyquist sampling theorem.

In summary, the duobinary filtering in the electrical domain reduces the total effective bandwidth of the PAM4 and thus permits the minimum distortion on the signals and the DSP can recover the original PAM signals with minimum reduction of the sensitivity of the receiver. Indeed this bit rate can be increased over 200 Gbps if the bandwidth of the system components can be enhanced to the upper limit of about 40–45 GHz. Alternatively, the OFDM by direction or the discrete multi-tone (DMT) technique can be employed to increase the transmission rate. This technique is demonstrated in the next few sub-sections. The disadvantage of OFDM is that the power consumption is quite high due to its high sampling rate. It is noted that the power dissipation is proportional to the cubic function of the sampling rate.

3.13 Beyond 1.0 Tbps capacity using IM/DD systems

It is possible to increase the bit rate to 150 G or even 200 G per channel so that with an aggregation of more optical carriers the total capacity per transmission line can reach more than 1.0 Tbps, as shown in figure 3.55. The system in the schematic of figure 3.55 and/or figure 3.49 needs to satisfy the following conditions: (i) the DAC/ADC bandwidth should be increased from 15 GHz to 25 GHz+; (ii) optical components should be increased to 30+ GHz; (iii) the driver and TIA should be increased to 30+ GHz; and (iv) the number of channels can be increased to 12 or even 16 from 8 optically modulated carriers, as shown in figure 3.56.

3.13.1 Beyond 200+ Gbps DMT transmission

The short-reach interface 400 GE[3] has been standardized and will be commercialized towards the year 2020, and is based on 100 Gbps/carrier. Currently, the next-generation 800 GE/1 TE (gigabit ethernet/terabit ethernet) has been seriously discussed with three possible solutions: coherent optics, eight-lane using 100/125 Gbps intensity modulation/direct detection (IM/DD), and four-lane using 200/250 Gbps IM/DD. Compared to the coherent solution as described in section 1.4 and in chapter 4, for a multi-carrier comb source, simple IM/DD configuration is more

[3] IEEE P802.3bs 400 Gb/s Ethernet Task Force, http://ieee802.org/3/bs/index.html.

attractive due to the advantages of complexity, cost and size. For IM/DD, systems with a smaller number of optical lanes (higher per-lane data rate) are strongly desired to reduce the complexity, size, power consumption and cost of the transceivers. Thus, a four-lane solution with beyond 200 Gbit s^{-1} per channel carrier is a promising candidate for 800 GbE/1 TE. 200 Gb s^{-1} PAM4 transmission has been reported as described in section 3.12, using a 40 GHz DAC and an in-house modulator with a bandwidth higher than 50 GHz [33]. Similarly, a 214 Gb s^{-1} PAM4 could be realized using an in-house electrical absorption modulator (EAM) with a bandwidth of 56 GHz [34]. Currently, the next-generation commercial DAC/ADC, driver and optical components (TOSA and ROSA) would not be higher than 35 GHz and possibly reach 45 GHz due to the electronic limitations of 65 nm analog micro-electronic fabrication technology. Thus a higher spectral efficiency format is required to achieve 200+ Gbps per lane (carrier) based on 35 GHz bandwidth class components. Discrete multi-tone (DMT) requires the narrowest bandwidth and has the highest spectral efficiency [35], which can be considered as an attractive candidate for an 800 GE/1 TE solution. A DMT with a gross data rate of 200 Gbps is realized, but without considering the FEC and other overheads [35]. A 250 Gb s^{-1} DMT is demonstrated by Yamzaki [36], however, two DACs are multiplexed to generate a high-bandwidth DAC, which increases the system's cost and complexity.

In this section, a low cost IM/DD DMT transmission system using a single wavelength, single DAC/ADC, single modulator and single PD is described. A new nonlinear equalization (NLE) is implemented based on a Volterra digital kernel filter upgraded by some absolute terms, so-called ABS-based NLE. The mean SNR is improved by 2.4 dB and 1 dB, compared to linear equalization and conventional NLE, respectively. In addition, enabled by trellis coding modulation (TCM), the BER performance is improved by more than one decade. Concerning a net rate of 200 Gb s^{-1}, 224 Gbps (7% HD-FEC is assumed) and 214 Gbps (KP4-FEC is assumed) data are successfully transmitted over a 1 km single-mode fiber (SMF) with BERs of 2.4e-4 and 2e-5, respectively.

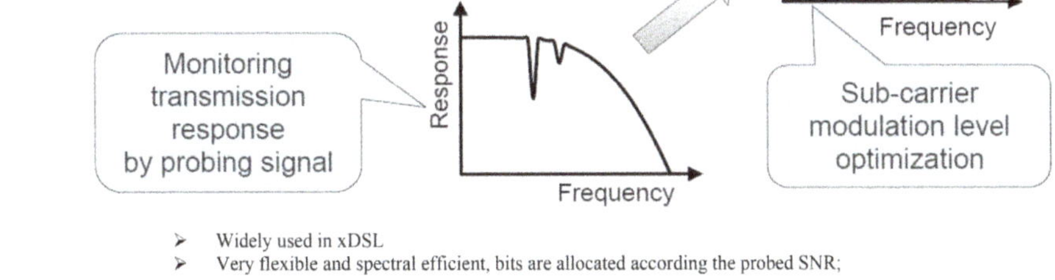

Figure 3.56. Principles of DMT modulation for extending the high to ultra-high capacity.

The operational principles of DMT are shown in figure 3.56. The optical carrier is modulated by electrical multi-carrier channels which are modulated in multi-level modulation formats and are orthogonal with each other. These electrically digitally modulated signals are generated in a DSP-based generator. Only the real parts of the complex signals are used as in the IM/DD any complex parts also result in real intensity. The spectrum of the multiple sub-carriers is shown in figure 3.56. These sub-carrier channels are shown after passing through a transfer curve of the electrical or optical components in figure 3.56 with an insert of the multi-QAM constellation of high and lower order. Optimization can be designed to accommodate the sensitivity of the receiving sub-section. The software architecture and parameters of the DMT system at the transmitter and reception sub-systems are shown in figure 3.57.

3.13.2 Experimental set-up

The sectional sub-routines are also stated, including the FEC. The experimental platform is shown in figure 3.58, and the parameters of the components of the

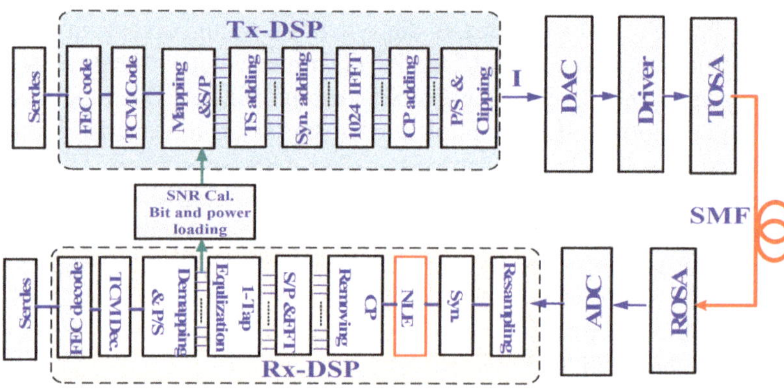

Figure 3.57. Algorithm processing architecture of DMT system at the transmitter section.

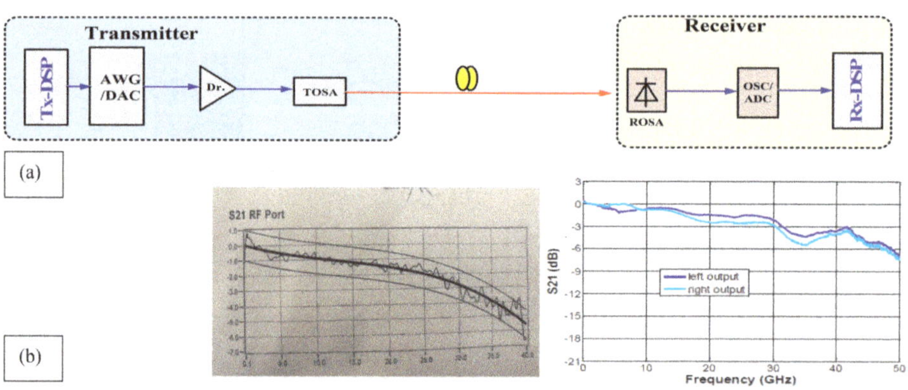

Figure 3.58. The beyond 200 G DMT platform set-up whose parameters shown in table 3.4.

sub-systems are given in table 3.4. The DSP is loaded with the generation of the DMT signals as described in figure 3.57. We experimentally prove the possibility of transmission beyond 200 Gbps employing DMT operational principles with the set-up shown in figure 3.57. The transmitter DSP is similar to the traditional DMT DSP except that the 16-state trellis coded modulation (TCM) encoder is employed to improve the system's performance. The IFFT size is 1024 and a cyclic prefix (CP) of 16 samples is added to alleviate the inter-symbol interference (ISI). The digital DMT data are converted to an analog signal by a Keysight arbitrary waveform generator (AWG (8196A)) with a 3 dB bandwidth of 30 GHz and a sampling rate of 92 GSa s^{-1}. The transmitter optical sub-assembly (TOSA) side consists of an SHF amplifier driver (807 s) which is used to boost the output of the AWG. The amplified electrical DMT signal with a V_{pp} of 2 V is employed to drive a Mach–Zenhder interferometric modulator (MZIM; Oclaro SD40). The bandwidth of the MZIM is 30 GHz and the V_{pi} is about 4.5 V. The frequency transfer functions of the optical modulator and electronic amplifier at the receiver are shown in the insets of figure 3.58 and are bandwidth limited to about 32–35 GHz.

In the receiver ROSA side, a 3 dB 50 GHz U^2t photodetector (PD (XPDV 2120R)) is employed to convert the optical to electrical signal. Since the PD has no trans-impedance amplifier (TIA) and the best input power is about 7 dBm, an optical pre-amplifier is employed to improve the sensitivity performance. A semi-conductor optical amplifier (SOA) is more suitable as an optical pre-amplifier with the advantages of low cost, low power consumption and ease of integration. Due to the lack of SOA in our laboratory, an erbium-doped fiber amplifier (EDFA) is used. A real-time oscilloscope (Keysight DSO X93304Q) is employed to capture the DMT signal at a sampling rate of 160 GSa s^{-1} and offline DSP processing is realized. In the Rx-DSP, the signal is first down-sampled from 160 GS s^{-1} to 92 GS s^{-1} to match the DAC sampling rate. Conventional NLE and ABS-based NLE are used to mitigate system nonlinearities including signal-to-signal beating interference (SSBI) caused by the direct detection process and modulator nonlinearity. After serial to parallel (S/P), CP removing and FFT, the signal is transformed to the frequency domain and one-tap equalization is used to compensate the linear distortions of the system. After de-mapping and P/S (parallel to serial) converting, a TCM decoder is used to demodulate the signal constellations to bit sequences using the Viterbi decoding algorithm.

Thus in summary, the aggregate 200 Gbps bit rate can be achieved via the multi-level modulation format imposed on the sub-carriers of the DMT which are varied in the high frequency region so as to satisfy the receiver sensitivity. Furthermore, the equalization by MLSE enables data recovery even in the high attenuation region of the transfer functions.

3.13.3 Performance

Figure 3.59(a) shows the SNR curves with a 3 dB bandwidth of about 23 GHz, which is mainly limited by the TOSA transmitter side. In the experiment, 508 sub-carriers are used that correspond to a bandwidth of 45.6 GHz. Without NLE, the mean SNR is 17.5 dB as shown by the blue line in figure 3.59. An improvement of

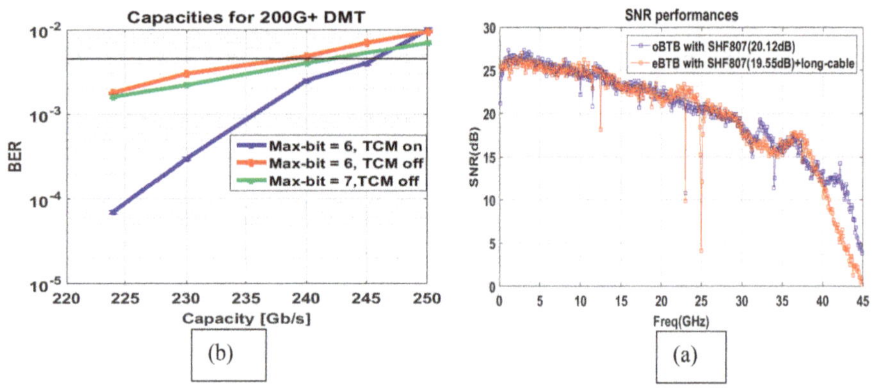

Figure 3.59. Optical B2B performance. B2B = back-to-back. (a) BER versus frequency with and without TCM. (b) SNR versus frequency with effects of variation of electrical coaxial cable.

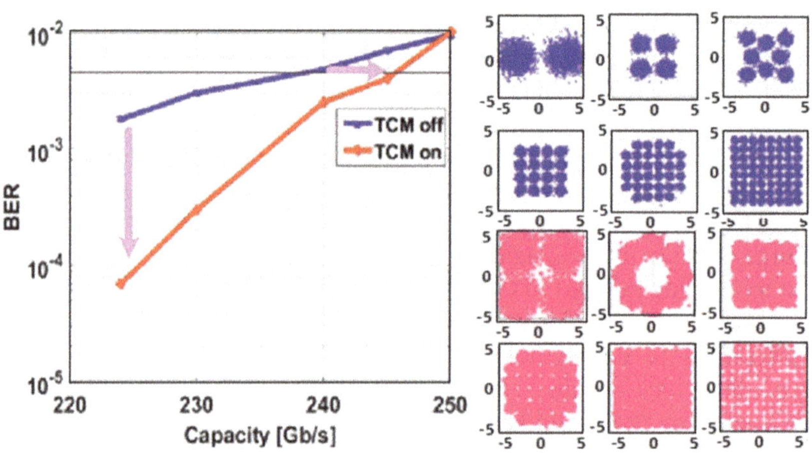

Figure 3.60. (a) B2B capacities with and without TCM; (b) constellations without (blue) and with (pink) TCM.

1.4 dB is achieved by a conventional second-order Volterra filter with terms x_i and $x_i x_j$, and dc. When two new terms $|x_i|x_i$ and $|x_i|x_i^2$ are introduced (ABS-based NLE), the mean SNR is further increased from 18.9 dB to 19.9 dB. In order to simplify the equalizer, the equalizer uses only three samples, meaning that only 10 and 16 taps are used in the conventional and ABS-based NLE modules, respectively.

We note the following. (i) TCM contributes a lot to the error-floor. (ii) For the TCM-on case, the best performance is 7e-5 for 224 G (200 G net rate); the maximum bit rate is 245 G @ 4e-3. (iii) For the TCM-off cases, the maximum of 7 bits (128QAM) brings some gain, thus we can ask whether a 7 bit modulation would bring some gain for the TCM-on case. The constellations of different modulation formats at different spectral regions of the transfer function of the sub-systems are shown in figure 3.60, operating under the B2B condition and under TCM-on or TCM-off scenarios. The bit-loading maps of 224 Gb s^{-1} and 214 Gb s^{-1} DMT are

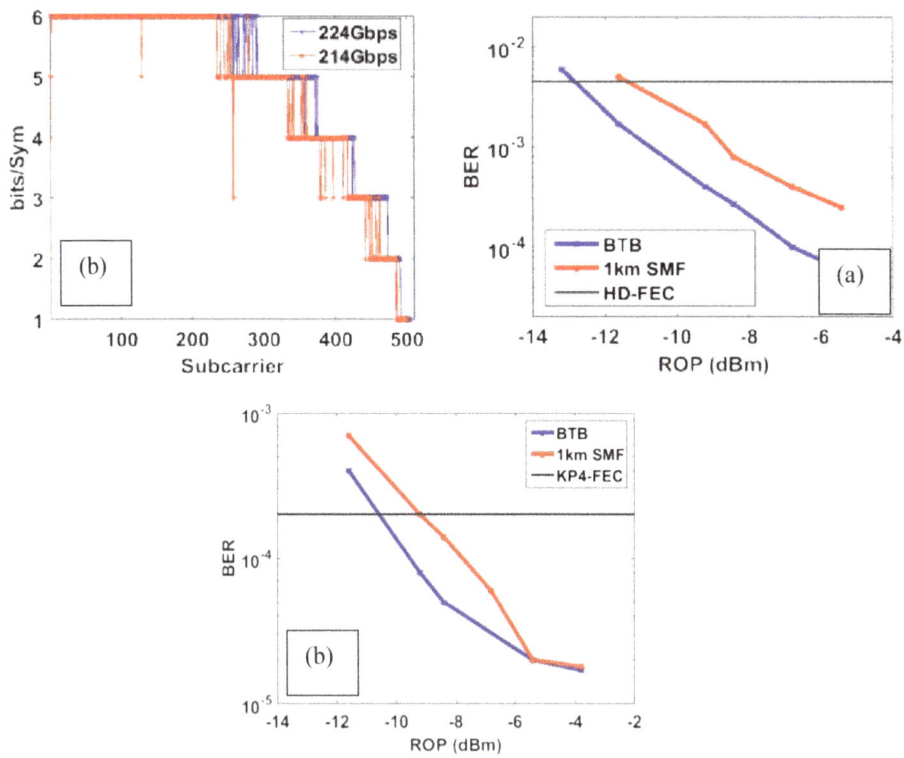

Figure 3.61. (a) Bit-loading maps of 224 Gb s^{-1} and 214 Gb s^{-1} DMT. (b) Power sensitivities of 224 Gb s^{-1} DMT. (c) Power sensitivities of 214 Gb s^{-1} DMT.

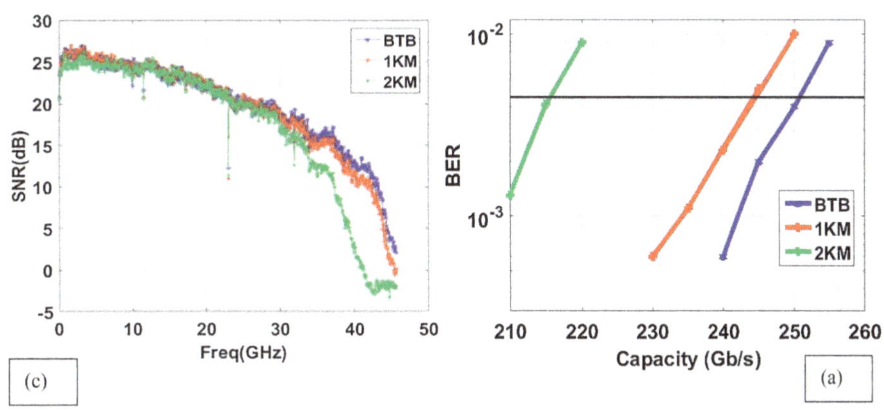

Figure 3.62. (a) SNR curves for B2B, 1 km and 2 km links. (b) Capacities in B2B, 1 km and 2 km cases.

shown in figure 3.61(a). Figure 3.61(b) shows the power sensitivities of 224 Gb s^{-1} DMT, and figure 3.61(c) the power sensitivities of 214 Gb s^{-1} DMT.

BER versus capacity (bit rate) for the B2B case with and without TCM is presented in figure 3.62(a). With time compression multiplex (TCM), not only is the capacity improved by 5 Gb s^{-1} (@ 4.5 × 10^{-3}), but also the BER performance is significantly improved from 2 × 10^{-3} to 7 × 10^{-5} (@ 224 Gb s^{-1}). The constellations with and without TCM are shown in figure 3.60. The TCM encoder adds one bit more and doubles the constellation points (e.g. increasing from QPSK to 8QAM). Although the constellation with TCM is a bit blurred, the TCM performance is improved due to the higher Euclidean distance by the convolutional encoder. A more complex TCM encoder can potentially increase the performance, however, in the following experiments the number of TCM states was fixed to 16 to lower the system complexity.

We note the following: (i) the transfer function of electrical B2B is not consistent with the SNR in high frequency; (ii) the noise is not white, so the Tektronix digital sampling oscilloscope could have introduced this type of noise; and (iii) the transfer function of the B2B is well matched with the SNR.

A net data rate of 200 GB s^{-1} was investigated with flexible FEC redundancy in different scenarios. For a high-performance (better power sensitivity) scenario, HD-FEC is employed with a gross data rate of 224 Gb s^{-1} with the DMT bit-loading map shown in figure 3.61(a) (blue line). Although 7 bits/symbol can be loaded in the low frequency spectral range, a maximum of 6 bits/symbol is mapped to enhance the system's noise tolerance. Figure 3.61(b) illustrates the receiver optical power (ROP) performance for the 224 Gb s^{-1} scenario with a sensitivity of −12.8 dBm in the B2B case, which is measured before the pre-amplifier. An approximately 1.2 dB power penalty is observed after 1 km C-band transmission, which is attributed to the 17 ps nm^{-1} km^{-1} chromatic dispersion (CD) parameter. The bit-loading map of 214 Gb s^{-1} is also shown in figure 3.61(a) (200 Gb s^{-1} net data rate and KP4-FEC), which is suitable for low-latency/power scenarios. Sensitivities of −10.4 dBm and −9.1 dBm are achieved for B2B and 1 km transmission, respectively. An error-floor at 1e-5 is observed, which is one decade below the FEC threshold and can be sufficient for real applications.

Figure 3.62(a) shows the SNR curves of B2B, 1 km and 2 km C-band transmissions, respectively. The SNR quickly decreases at frequencies above 30 GHz in the 2 km link, which is due to the accumulated CD (~34 ps nm^{-1} by 2 km SSMF). To study the maximum system capacity, we vary the data rate and record the BER performance. Transmissions of 250 Gb s^{-1}, 244 Gb s^{-1} and 216 Gb s^{-1} over 0 km, 1 km and 2 km SMF with a BER below pre-FEC of 4.5e-3 (HD-FEC), as shown in figure 3.62(b), are successfully demonstrated, with a net data rate of 223 Gb s^{-1}, 218 Gb s^{-1} and 193 Gb s^{-1}. Since SSMF operating in the C-band exhibits a much higher CD than that in the O-band, we believe that our solution can be used for 2 km or even 10 km O-band transmissions with negligible power and capacity penalties (figure 3.63).

Capacities for 200+ G per carrier can be seen, as depicted in figure 3.64. Aggregation of many optical modulated channels can offer multi-Tbps total capacity, as shown in figure 3.65. The optics and signal processing optical receivers for such aggregation are shown.

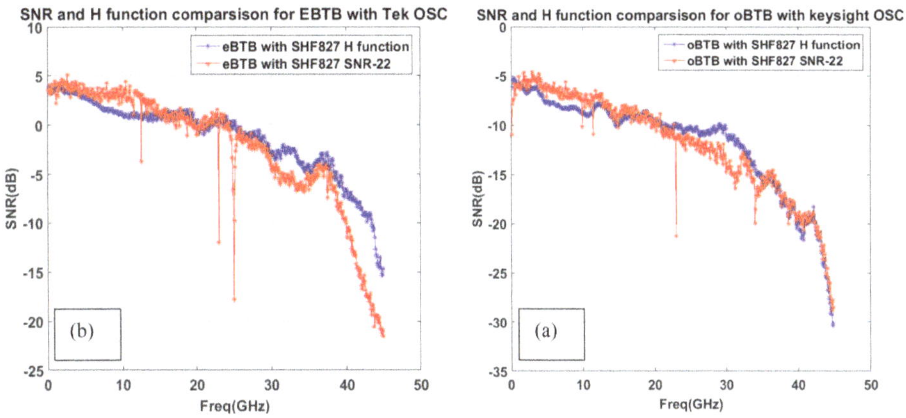

Figure 3.63. DMT transmission system: SNR versus frequency with H-function.

3.13.4 Remarks

A beyond 200 Gbps DMT system is demonstrated with a simple architecture. Based on the proposed ABS-based NLE and TCM, the system's sensitivity performance and error-floor are significantly improved. With the help of an optical pre-amplifier, receiver sensitivities of −12.8 dBm and −10.4 dBm are achieved for hardware forward error coding (HD-FEC) and class protocol 4 forward error coding (KP4-FEC), which currently is the best sensitivity performance for a 200 Gbps IM/DD system. The experimental results show that our 200 Gb s^{-1} DMT scheme is a very promising low cost and high capacity candidate for the next generation of 800 GE/ 1 TE systems using four-lane optics. GE = gigabit ethernet.

3.14 Higher order modulation in IM/DD systems

Following the standardized short-reach 400 gigabit ethernet (GE)[4], the next-generation 800 GE or 1.6 terabit ethernet (TE) is needed [37] beyond 2020. For short-reach applications below 2 km, intensity modulation and direct detection (IM/DD) solutions are preferred to coherent solutions due to their lower complexity, cost, size and power consumption, with somewhat lower sensitivity. Compared to the 100 G/lane approach, the 200 G/lane solutions are widely viewed as more compact and cheaper [38–41]. Recently Wei *et al* have conducted an experimental demonstration for comparison of 200 G per lane of different higher order modulation formats under IM/DD techniques. The system is shown in figure 3.66 in which DSPs are used in both the transmitting and receiving sub-systems so as to generate via software a flexible modulation format in these sub-systems. This flexibility allows users to increase or decrease the capacity in such links whenever desired. The feasibility of 200 G/λ short-reach DCN based on 30 GHz components[1] [5–9, 12] has been shown using advanced modulation formats and DSP. Demonstrations mainly considered single carrier pulse amplitude modulation (PAM) [4–6, 8–12] and multi-carrier DMT schemes [3, 7, 10]. Trellis coded modulation

[4] IEEE P802.3bs 400 Gb/s Ethernet Task Force, 2018. http://ieee802.org/3/bs/.

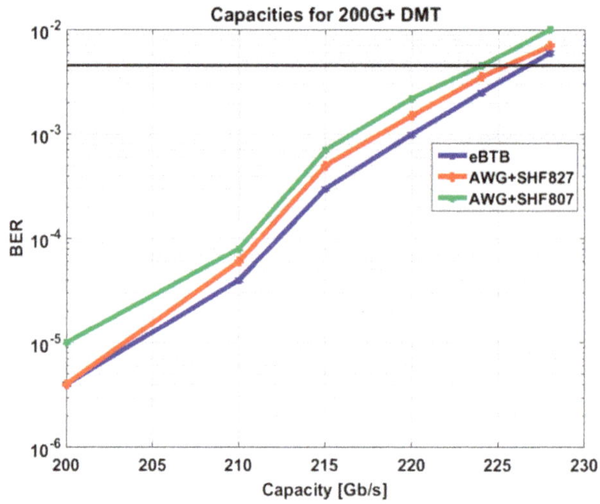

Figure 3.64. Capacities for 200+ G employing the components as listed in table 3.5 and variation of the RF drivers.

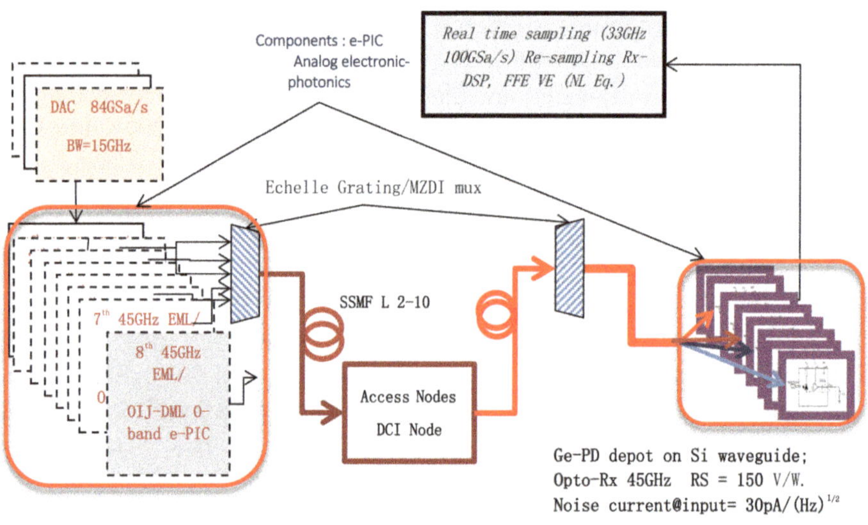

Figure 3.65. Multi-carrier multichannel to aggregate transmission capacity to beyond Tbps (8 × 150 Gbps giving a total of 1.2 Tbps). Multi-lane (8×) transceiver e-PIC system set-up. Transceiver: analog e-PIC in association with DSP. RS = receiver sensitivity in *V*/*W*, that is for an optical power of one watt the receiver gives 150 V voltage output. So for a typical average input power of 10 μW the output of the receiver is about 1.5 mV entering the ADC. OJ = optical injection; DML = direct modulated laser.

(TCM), which is a promising approach in the presence of bandwidth and noise limitations, has been used in both single carrier and multi-carrier systems [7–9].

The choice of modulation techniques influences the trade-off between transmission system performance and processing complexity due to cost and power factors. Experimental comparison of various advanced IMDD modulation formats,

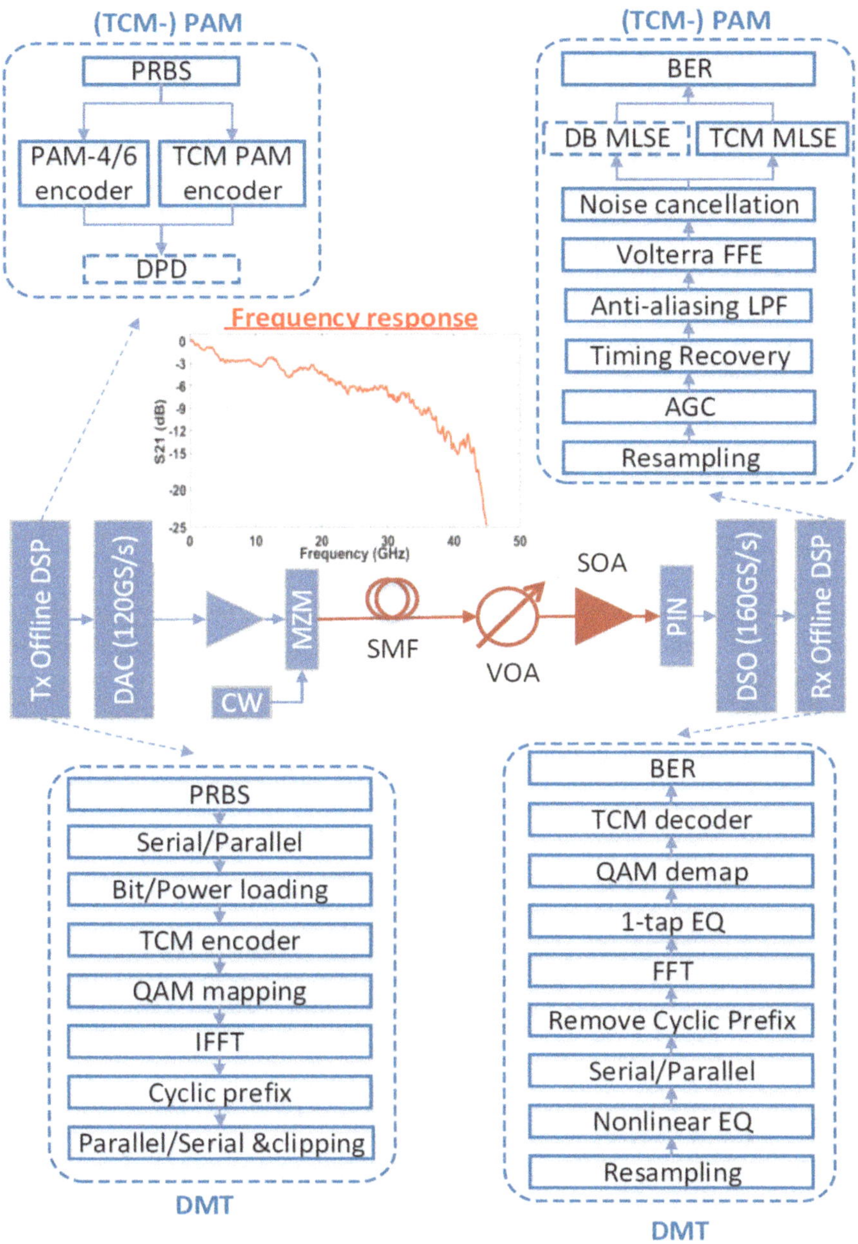

Figure 3.66. Schematic of the experimental set-up for 200 G systems.

including duobinary (DB) PAM4, PAM6, two-dimensional (2D)-PAM8 TCM and DMT for DCN applications, has demonstrated the flexibility for capacity exchanges in these short distance transmissions, in particular in data center interconnection (DCI) or short distance access. The DCI technology is based on band limited components to 30 GHz to achieve a 200 Gbt s^{-1} rate or an effective 7 bits Hz^{-1} IM/DD under DMT modulation format and PAM4 or PAM6.

This demonstration is performed in the C-band with EDFA. The transmitter loads a soft DSP to generate the digital waveforms via a 120 GSa s^{-1} 30 GHz bandwidth arbitrary waveform generator (AWG) to convert them into an analog electrical signal. A linear electronic driver with 50 GHz 3 dB bandwidth is used to single-drive a 33 GHz 3 dB Mach–Zehnder intensity modulator (MZIM) coupled with a single-mode laser operating at 1550 nm, an SOA pre-amplified optical receiver plus a variable optical attenuator (VOA) and a 60 GHz photodiode (PD). The SOA gain is adjusted to optimize the input power of the PD. The detected signal is sampled and digitally processed in a 160 GSa s^{-1} 60 GHz 3 dB bandwidth real-time sampling oscilloscope. The end-to-end (from AWG to oscilloscope) frequency response of the optical B2B link is shown in the inset of figure 3.66, indicating an overall effective 3 dB bandwidth of only 19 GHz. The digital signal is processed by offline DSP. The detailed offline DSP for the transmitter and the receiver is also presented in figure 3.66 for all modulation formats of PAM, TCM PAM, as well as DMT schemes.

For both DB-PAM4 and PAM6, the transmitter DSP includes a PRBS generator, a PAM bit-mapper. Furthermore a linear 16-tap digital pre-distortion (DPD) with re-sampling is employed at the transmitting sub-system to compensate for the limited overall bandwidth.

For PAM6, the encoder maps the input bits to a 32QAM constellation, whose I and Q components are transmitted over two consecutive symbols [41, 42]. The DPD aims to compensate the limited bandwidth of the transceiver [5] and can be optionally switched on or off. In the offline DSP at the receiver, the signal is re-sampled to two samples per symbol. An automatic gain control (AGC) followed by a timing recovery (TR) and an anti-aliasing low-pass filter is then applied. The signal is then equalized by a symbol-spaced Volterra filter, which contains linear, second- and third-order kernels. A DB or a non-DB feed forward equalization (FFE) is applied, depending on the trade-off between noise enhancement and SNR requirements [41]. A following noise cancellation (NC) unit is used to suppress and whiten the additive noise [43]. If the FFE is trained to deliver a non-DB signal, the subsequent processing consists only of a decision device before the calculation of the bit error rate (BER). Otherwise, in the case of DB-FFE, a DB maximum likelihood sequence estimator (MLSE) [42] is used to decode the DB PAM signal.

2D-PAM8 employs a similar processing to the PAM4 and PAM6 mentioned above with an incorporation of a convolutional TCM encoder and a TCM MLSE decoder in the transmitter and the receiver, respectively.

Figure 3.67 depicts the overall bandwidth of the channel launching into the transmission line and after the reception sub-systems.

The receiver sensitivity of the 200 Gbit s^{-1} (plus overhead rate) of the PAM4 and PAM6 modulation formats under different processing techniques is shown with BER in figure 3.68 of the B2B and 1.0 km SSMF transmission link. Table 3.5 summarizes the effectiveness of FEC/DSP for the higher modulation formats on the system performance.

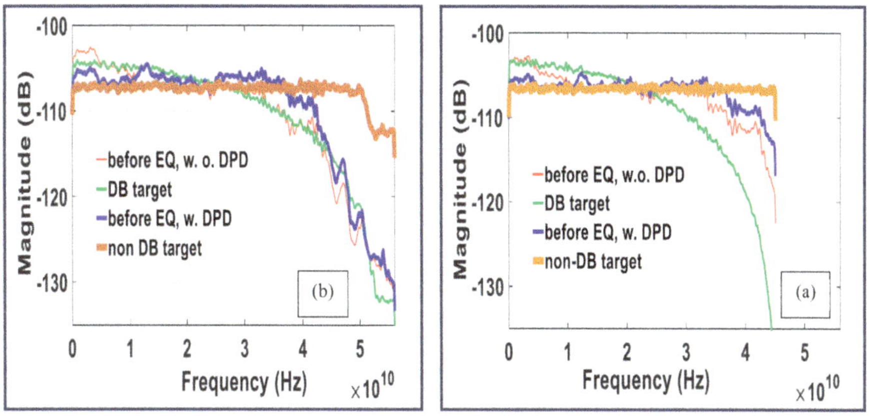

Figure 3.67. (a) 112 GBd PAM4 and (b) 90 GBd PAM6 signal spectra before and after receiver FFE. EQ = equalizer; DPD = detection photodetector; DB; duobinary.

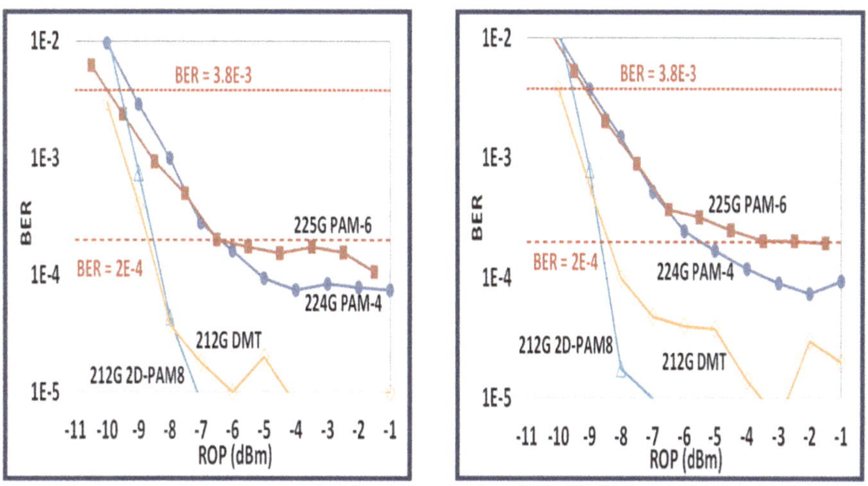

(a) B2B transmission (b) 1.0Km SSMF transmission

Figure 3.68. BER versus receiver sensitivity (minimum power at receiver) of each scheme for the (a) optical B2B and (b) 1 km SSMF cases. Under different higher order modulation as indicated.

In the above experimental comparison of the most promising 200 G/lane or (λ) DCN IM/DD techniques, employing band-limited 30 GHz components, DB-PAM4 and PAM6 are identified to offer very effective links with high sensitivity under PAM6 with simpler processing power. PAM4 would be an attractive solution if wider passband components are available. 2D-PAM8 allows the use of simpler FEC but requires relatively high power processing DSP.

3.15 Concluding remarks

The explosive growth of traffic in DCNs is driving the demand for high capacity short-reach interconnections. Following the standardized short-reach 400 gigabit

Table 3.5. Performance and FEC/DSP algorithms for 200 G under different modulation formats.

Format	224 G DB-PAM4	225 G PAM6	212 G 2D-PAM8	212 G DMT
1 km ROP sensitivity	−9 dBm	−9 dBm	−8.6 dBm	−8.4 dBm
FEC overhead (%)	12%	12%	5.9%	5.9%
DSP pocessing/equalization	DPD, DB-FFE, DB-MLSE	DPD, FFE	DPD, FFE, TCM	IFFT/FFT, FFE, TCM
1 km ROP sensitivity	−9 dBm	−9 dBm	−8.6 dBm	−8.4 dBm
FEC overhead (%)	12%	12%	5.9%	5.9%
DSP and equalization	DPD, DB-FFE, DB-MLSE	DPD, FFE	DPD, FFE, TCM	IFFT/FFT, FFE, TCM, IFFT/FFT

ethernet (GE)[3], the next-generation 800 GE or 1.6 terabit and then exabit ethernet (tera-E/exa-E) are essential and urgently needed [37] beyond 2020 when 5G network technologies will already be deployed. For short-reach applications below 2 km, typically in DCN interconnection and ultra-high capacity access nodes, intensity modulation and direct detection (IMDD) solutions are preferred as well as coherent reception solutions. IM/DD are sometimes preferred because of their lower complexity, cost and size. However, unlike the coherent transmission technique in which both the polarization modes and the quadrature phase components can be used to increase the transmission capacity by a factor of at least 4, the IM/DD can only be employed at a higher order of modulation and the processing power to achieve ultra-high bit rate, in particular in the effectiveness of the sensitivity and equalization. Compared with the 100 G/lane approach, 200 G per-lane solutions are widely viewed as more compact and cheaper. However, currently there is a lack of commercial components that can satisfy the requirements for bandwidth and SNR. For this reason, a few demonstrations showing remarkable performance experimental high-bandwidth components, such as a special high speed DAC, have been outlined and described in this chapter. Even with high-performance components, powerful DSP is required to mitigate linear and nonlinear distortions. This chapter has also outlined the coherent reception techniques that we expect for the coherent sub-systems with low cost Si photonic components that will be available in the near future.

The high capacity transmission techniques described in this chapter allow us to understand the main principles and specific applications for DCI networking.

3.16 Appendix 1: Principles of DSP-based coherent transmission

For DSP-based coherent transmission systems, both the transmitter and receiver require DACs, ADCs and DSPs so that DSP can be employed in association with

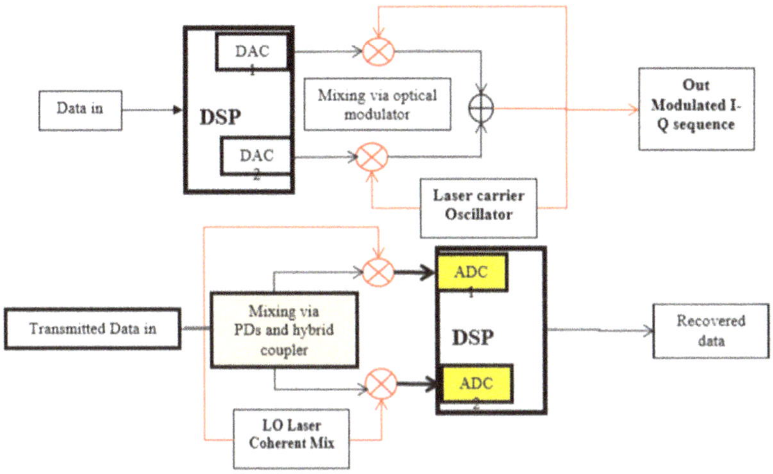

Figure 3.69. Schematic of coherent transmitter and coherent reception systems for coherent transmission systems.

algorithms to recover the transmitted data sequences of both polarized channels. These DSP systems are the distinct difference in contrast to the first coherent systems developed in the 1980s [44]. The optical channels are polarization multiplexed and modulated with the in-phase and quadrature signals under appropriate biasing points so that complex QAM signals can be generated. In the reception sub-system the transmitted channels are mixed with an optical local laser via a hybrid coupler so that the polarized channels can be demultiplexed and even the in-phase and quadrature components can be split. These optical domain channels are then detected via balanced reception and electronically amplified and sampled by ADCs and thence processed in the DSP systems (figure 3.69). More details of these reception systems are described in sections 1.2 and 3.3.

3.17 Appendix 2: Balanced detection in coherent receivers

The quality of balanced detection in a coherent receiver is very important due to the non-ideal condition of the two photodetectors connected B2B. The impact of this imbalance generates distortion which can then be defined by the term common mode rejection ratio (CMRR). Furthermore, the optical power split and polarization diversity in the $\pi/2$ hybrid coupler has significant impact on the optical receiver. Thus the impact of the characteristics of the optics on the balanced detection is analyzed. Parameters that characterize the performance of the balanced detection suitable for the optical front end (OFE) are proposed.

3.17.1 Optical front end

The structure of an OFE using polarization diversity and coherent detection is shown in figure 3.70. The structure consists of two hybrid couplers for the X- and Y-polarization channels of the transmitted data streams. The local oscillator is also

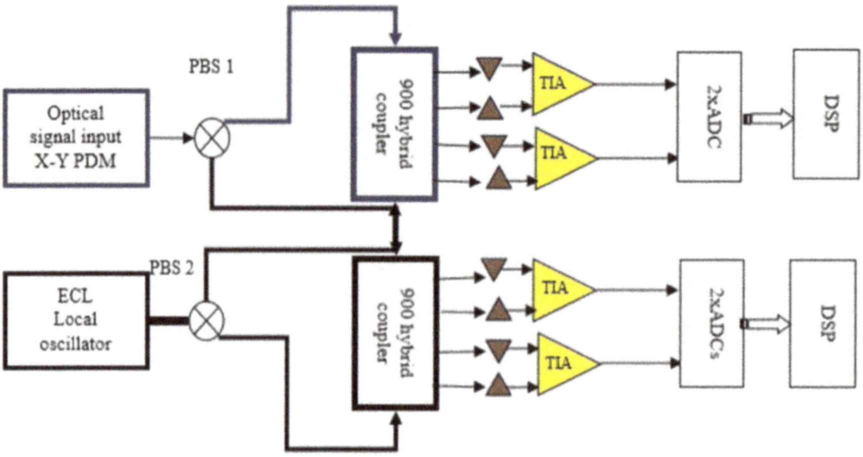

Figure 3.70. Block diagram of a digital coherent optical receiver (OFE) using a PD pair and differential amplifier or two-port optical receiver. TIA = trans-impedance amplifier with a transfer resistance (current input giving output voltage) in the mid-band or a few kilo-ohms.

polarization split and fed into the coupler so that mixing between the oscillator and the signal polarized modes can occur.

3.17.2 Optical mixing and polarization diversity

Optical signals and local oscillators are split in terms of polarization and power, and their mutual mixing happens in the $\pi/2$ hybrid coupler. The 900 optical section ensures that the distinction between the in-phase (I) and quadrature phase (Q) components of the optical signal can be distinguished.

Total insertion loss per polarization would be about 10 dB including the 3 dB power splitting to the two polarized channels. For each channel of I and Q there are two complementary outputs in the optical domain where the phase difference would be π. Thus a balanced PDP (photodetection pair) would give an extra 3 dB gain in the common mode and be depleted in the difference mode.

3.17.3 Differential amplification

The differential detection and amplification of optical signals (now PD currents) is normally done using a single input port with the two PDs connected B2B and then fed into a trans-impedance amplifier as shown in figure 3.71(a). Alternatively the photocurrents can be fed separately into two differential inputs of a differential trans-impedance amplifier, as shown in figure 3.71(b).

The input fields fed into the two PDs can be written as

$$E_s(t) = \sqrt{P_s(t)}\, e^{j\omega_s t} e^{j\phi_s(t)} \tag{3.11}$$

$$E_{\mathrm{LO}}(t) = \sqrt{P_{\mathrm{LO}}(t)}\, e^{j\omega_{\mathrm{LO}} t} e^{j\phi_{\mathrm{LO}}(t)}. \tag{3.12}$$

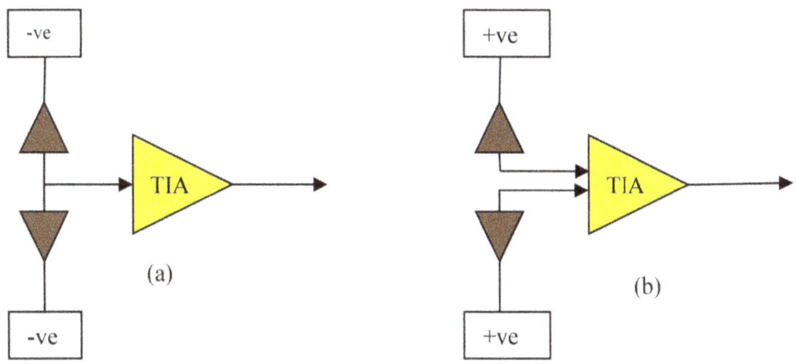

Figure 3.71. Arrangement of a PD pair for balanced detection: (a) a single-end trans-impedance amplifier (TIA) and (b) a differential balanced amplifier.

Thus after mixing and polarization splitting, the detected currents at the outputs of the PDP are given by

$$i_1(t) = \frac{1}{2}\Re\left\{P_s(t) + P_{\mathrm{LO}s}(t) + 2\sqrt{P_{\mathrm{LO}}P_s(t)}\,\sin(\omega_s - \omega_{\mathrm{LO}})t + \phi_s(t) - \phi_{\mathrm{LO}}\right\} \quad (3.13)$$

$$i_2(t) = \frac{1}{2}\Re\left\{P_s(t) + P_{\mathrm{LO}s}(t) - 2\sqrt{P_{\mathrm{LO}}P_s(t)}\,\sin(\omega_s - \omega_{\mathrm{LO}})t + \phi_s(t) - \phi_{\mathrm{LO}}\right\}. \quad (3.14)$$

In equations (3.13) and (3.14) there are three terms: (i) photocurrents produced by the signal itself; (ii) currents produced by the power of the local oscillator; and (iii) the useful signal current whose amplitude is boosted by the local oscillator power. The first current due to signal happens at a frequency far away from the spectral sensitivity of the PD material, the second term produces a dc photocurrent and then turns into a shot noise current which creates more noise in the signal output. The third term is a useful term that represents the phases of the signals in the I and Q components.

3.17.4 Unmatched detector frequency responses

A further difficulty that may degrade the performance of a dual-detector receiver occurs when the two detectors have differing frequency responses. This problem could arise, for example, if the detectors have different bandwidths. For good cancellation, the LO noise photocurrents in both detectors must be well matched both in amplitude and phase over the entire frequency band of interest. If one detector begins to roll off before the other near the edge of the frequency band, poor cancellation could result. For 20 dB of cancellation with a 50:50 beam splitter and the two attenuators equal to 1, the detector responses must be matched within at least 1.6 dB (assuming matched phases) and 11‰ (assuming matched amplitudes). Clearly, if one detector has rolled off by 3 dB and acquired a 45° phase shift relative to the other, poor cancellation will result. Unmatched detector frequency responses can also occur because of line resonances in the two signal paths when the detectors are physically connected into a circuit. If the noise in each detector is unequally amplified due to differing line resonances in the two signal paths, the balanced mixer action of the dual-detector receiver may be disrupted. In fact, an attenuator designed to maximize the signal-to-noise differential ratio (SNDR) of a particular receiver might change the line resonances of the signal path in which it is placed enough to lower the overall SNDR rather than increase it as predicted (figure 3.72).

3.18 Intra-DC networking and access transmission

The generic intra-DC interconnection is shown on the left-hand side of figure 3.73 with relevant growth of the market related to this sector on the upper right-hand side corresponding to its capacity growth.

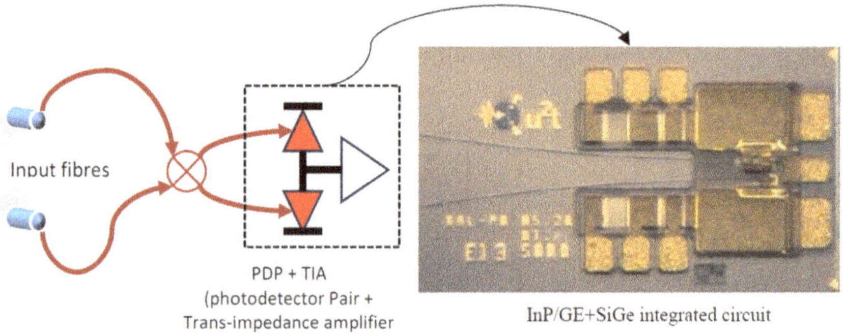

Figure 3.72. PDP differential amplifier optical receiver: (a) schematic diagram and (b) realized integrated circuit with two guided wave structures.

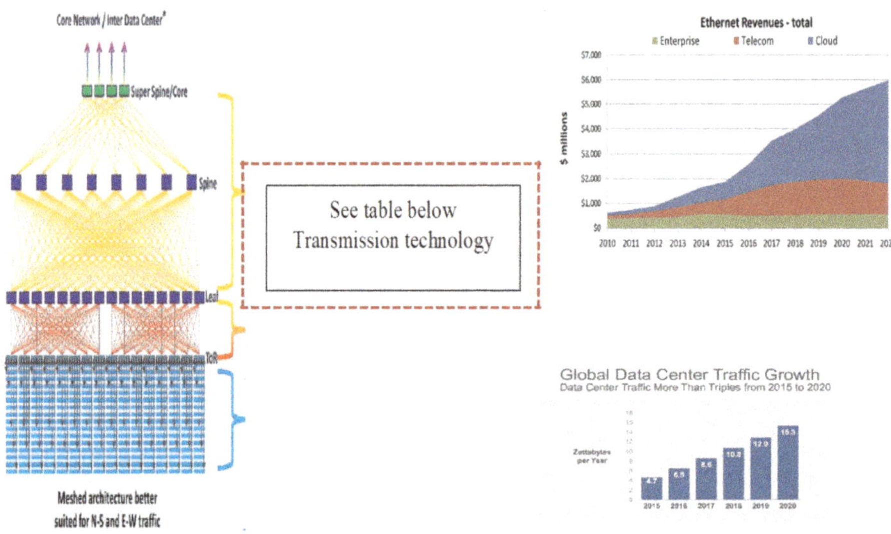

Figure 3.73. Data center interconnections and growth strength.

Table 3.6. 500 m to 2 km transmission link for intra-DC networking—current and future technology. DCI = data center interconnection.

Application	Transmission distance	Port ratio	Current technology	Near future
DCI	10 km	25%	Coherent 400 ZR	SMF
Metro	80 km		80K km	Coherent/DD
Spine ←→ Core	500 m–2 km	25%	CWDM4, 100 G/400 G, SMF	SMF, DD
TOR ←→ Leaf	100–300 m	50%	400 G, MMF	
Leaf ←→ Spine			8 parallel	
Server ←→ TOR	⩽30 m	AOC	25 G	

Table 3.6 summarizes the transmission technique and distance for 500 m to 2 km for intra-DC communications. Both incoherent and coherent reception are outlined. Although coherent reception might be a bit more expensive, it is expected to surpass direct detection due to its high sensitivity, thus no optical amplifier and DSP recovery need to be used.

References

[1] Soref R A and Lorenzo J P 1986 All-silicon active and passive guided-wave components for $\lambda = 1.3$ and 1.6 microns *IEEE J. Quantum Electron.* **22** 873–9

[2] Jalali B and Fathpour S 2006 Silicon photonics *IEEE J. Lightwave Technol.* **24** 4600–15

[3] Kikuchi K 2016 Fundamentals of coherent optical fiber communications *IEEE J. Lightwave Technol.* **34** 157

[4] Bell T E 1986 Communications: coherent optical communication shows promise, the FCC continues on its path of deregulation, and satellite communications go high-frequency *IEEE Spectr.* **23** 49–52

[5] Savory S J 2010 Digital coherent optical receivers: algorithms and subsystems *J. Sel. Top. Quantum Electron.* **16** 1164–78

[6] Pfau T, Hoffmann S and Noe R 2009 Hardware efficient digital receiver concept with feed forward carrier recovery for M-QAM constellation *IEEE J. Lightwave Technol.* **27** 989–99

[7] Mendinueta J M D, Bayvel P and Thomsen B C 2011 Digital lightwave receivers: an experimentally validated system model *IEEE Photonics Technol. Lett.* **23** 338–40

[8] Zhou X and Yu J 2009 200-Gb/s PDM-16QAM generation using a new synthesizing method *ECOC2009* paper 10.3.5

[9] Gnauck A H and Winzer P J 2009 10 × 112-Gb/s PDM 16-QAM transmission over 1022 km of SSMF with a spectral efficiency of 4.1 b/s/Hz and no optical filtering *ECOC2009* paper 8.4.2

[10] Stojanovic N 2008 An algorithm for AGC optimization in MLSE dispersion compensation optical receivers *IEEE Trans. Circuits Syst.* I **55** 2841–7

[11] Mao B, Stojanovic N, Xie C, Chen M, Binh L and Yang N 2011 Impacts of ENOB on the performance of 112 Gbps PDM-QPSK *ECOC2011* paper We.10.P1.46

[12] Kester W 2006 ADC input noise: the good, the bad, and the ugly. Is no noise good noise? *Analog Dialogue* **40–01** 13–7

[13] Regenbogen L K 1980 Nonlinearity model with variable knee sharpness *IEEE Trans. Aerosp. Electron. Syst.* **16** 874–7

[14] Weiner J S *et al* 2003 SiGe differential transimpedance amplifier with 50 GHz bandwidth *IEEE J. Solid-State Circuits* **38** 1512–7

[15] Ip E and Kahn J M 2008 Compensation of dispersion and nonlinear impairments using digital back propagation *IEEE J. Lightwave Technol.* **26** 3416

[16] Fatadin I, Ives D and Savory S J 2010 Laser linewidth tolerance for 16-QAM coherent optical systems using QPSK partitioning *IEEE Photonics Technol. Lett.* **22** 631–3

[17] Louchet H, Kuzmin K and Richter A 2008 Improved DSP algorithms for coherent 16-QAM transmission *Proc. ECOC'08 (Belgium)* paper tu.1.E6

[18] Noe R, Pfau T, El-Darawy M and Hoffmann S 2010 Electronic polarization control algorithms for coherent optical transmission *IEEE J. Sel. Quantum Electron.* **16** 1193–9

[19] Binh L N 2010 Technical specification of corning fiber G.652 SSMF *Digital Optical Communications* (Boca Raton, FL: CRC Press) chapter 3, appendix

[20] Godard N 1978 Passand timing recovery in an all-digital modem receiver *IEEE Trans. Commun.* **26** 517–23

[21] Stojanovic N, Gonzalez N G, Xie C, Zhao Y, Mao B, Qi J and Binh L N 2012 Timing recovery in Nyquist coherent optical systems *Int. Conf. Telecommunications Systems (Serbia)*

[22] Fang T T and Liu C F 1993 Fourth-power law clock recovery with pre-filtering *Proc. ICC (Geneva, Switzerland) vol 2 pp 811–5*

[23] Dong Z, Li X, Yu J and Chi N 2012 128-Gb/s Nyquist-WDM PDM-16QAM generation and transmission over 1200-km SMF-28 with SE of 7.47 b/s/Hz *IEEE J. Lightwave Technol.* **30** 4000–6

[25] Zhou X, Nelson L E, Magill P, Isaac R, Zhu B Y, Peckham D W, Borel P I and Carlson K 2012 PDM-Nyquist-32QAM for 450-Gb/s per-channel WDM transmission on the 50 GHz ITU-T grid *IEEE J. Lightwave Technol.* **30** 553

[26] Zhou X, Nelsonl L E, Isaac R, Magill P, Zhu B, Peckham D W, Borel P and Carlson K 2012 1200 km transmission of 50 GHz spaced, 504-Gb/s PDM-32-64 hybrid QAM using electrical and optical spectral shaping *OFC/NFOEC, (4-8 March 2012, Los Angeles, CA)*

[27] Bosco G, Curri V, Carena A, Poggiolini P and Forghieri F 2011 On the performance of Nyquist-WDM terabit superchannels based on PM-BPSK, PM-QPSK, PM-8QAM or PM-16QAM subcarriers *IEEE J. Lightwave Technol.* **29** 53

[28] Zhu R, Xu K, Zhang Y, Li Y, Wu J, Hong X and Lin J 2008 QAM coherent subcarrier multiplexing system based on heterodyne detection using intermediate frequency carrier modulation *2008 Microw. Photon.* pp 165–8

[29] Nakazawa M, Yoshida M, Kasai K and Hongou J 2006 20 Msymbol/s, 64 and 128 QAM coherent optical transmission over 525 km using heterodyne detection with frequency-stabilized laser *Electron. Lett.* **42** 710–2

[30] Dong Z, Li X, Yu J and Yu J 2013 Generation and transmission of 8 × 112-Gb/s WDM PDM-16QAM on a 25-GHz grid with simplified heterodyne detection *Opt. Express* **21** 1773

[31] Luo S *et al* 2015 Graphene-based optical modulators *Nanoscale Res. Lett.* **10** 199

[32] Patel D, Parvizi M, Ben-Hamida N M, Rolland C and Plant D V 2018 Frequency response of dual-drive silicon photonic modulators with coupling between electrodes *Opt. Express* **26** 8409–15

[33] Binh L N 2009 *Digital Optical Communications* (Boca Raton, FL: CRC Press, Taylor and Francis) chapter 9

[34] Kanazawa S *et al* 2016 Transmission of 214-Gbit/s 4-PAM signal using an ultra-broadband lumped-electrode EADFB laser module *Proc. OFC* paper Th5B.3
[35] Lange S *et al* 2018 100 GBd intensity modulation and direct detection with an InP-based monolithic DFB laser Mach–Zehnder modulator *J. Lightwave Technol.* **36** 97–102
[36] Zhang L *et al* 2018 Nonlinearity-aware 200 Gbit/s DMT transmission for C-band short-reach optical interconnects with a single packaged electro-absorption modulated laser *Opt. Lett.* **43** 182–5
[37] Yamazaki H *et al* 2016 300-Gbps discrete multi-tone transmission using digital-preprocessed analog-multiplexed DAC with halved clock frequency and suppressed image *Proc. ECOC* paper Th3B.4
[38] Wei J *et al* 2015 400 Gigabit ethernet using advanced modulation formats: performance, complexity, and power dissipation *IEEE Commun. Mag.* **52** 182–9
[39] Yamazaki H *et al* 2018 Transmission of 400-Gbps discrete multi-tone signal using >100-GHz-bandwidth analog multiplexer and InP Mach–Zehnder modulator *Proc. ECOC (Rome, Italy)* pp 1–3
[40] Estarán J M *et al* 2019 140/180/204-Gbaud OOK transceiver for inter- and intra-data center connectivity *J. Lightwave Technol.* **37** 178–87
[41] Zhang F *et al* 2019 Up to single lane 200 G optical interconnects with silicon photonic modulator *Proc. OFC (San Diego, CA)* pp 1–3 paper Th4A.6
[42] Wei J *et al* 2018 Linear pre-equalization techniques for short reach single lambda 225 Gb/s PAM IMDD systems *Proc. ECOC (Rome, Italy)* pp 1–3
[43] Stojanović N *et al* 2019 4D PAM-7 Trellis coded modulation for data centers *IEEE Photonics Technol. Lett.* **31** 369–72
[44] Wei J, Stojanović N and Xie C 2018 Nonlinearity mitigation of intensity modulation and coherent detection systems *Opt. Lett.* **43** 3148–51
[45] Li G 2009 Recent advances in coherent optical communication *Adv. Opt. Photonics* **1** 279–307
[46] Verbist J *et al* 2017 DAC-less and DSP-free PAM-4 transmitter at 112 Gb/s with two parallel GeSi electro-absorption modulators *Proc. ECOC 2017 (Gothenburg)*

Chapter 4

Superchannel transmission by multi-carrier sources

Transmission with ultra-high capacity for interconnection of data centers (DCs), in networking cores and networking within DCs is critical for the zetta-bits s^{-1} global Internet[1]. The low cost aspect is important to lower the total cost per user. Thus for multi-channels carrying 100 G and beyond, the employment of a multi-carrier source would drastically reduce the cost compared to if several single sources are used. This chapter presents the stable generation of comb lasers by re-circulating a fiber loop incorporating integrated synchronous optical modulators (ISOMs). The ISOM is composed of a master and slave configuration of Mach–Zehnder waveguide interferometers operating in such a mode that the optical phase paths are in synchronization. The generated sub-carrier spectrum can be frequency shifted to the left or to the right side or to both sides of the main optical carrier without suffering from the interference of the other sideband shifted sub-carriers by applying Hilbert transformed electric oscillation driving signals to the dual electrodes of the ISOM, hence suppression of one shifted band. The generation of sub-carriers for the left-side or right side shifting and the dual band spectrum are experimentally demonstrated. Single sideband (SSB) comb generators are then incorporated in tera-bits s^{-1} (Tbps) optical transmission systems and demonstrated in an installed fiber network, allowing the transmission of Tb s^{-1} superchannels over long distance non-compensating optically amplified lines and optical coherent receivers incorporating digital signal processors (DSPs). Nyquist pulse shaping of data sequences are implemented using a digital-to-analog converter (DAC) with 56 to 64 GSa s^{-1} so as to achieve effective spectrum packing of the modulated sub-carrier channels. The obtained transmission performance (BER versus OSNR) of 10^{-3} with OSNR of 16 dB shows that the sub-carriers of

[1] Zetta = 10^{+18}.

doi:10.1088/978-0-7503-2292-8ch4

the SSB comb generator can provide a stable source for multi-Tbps transmission systems. It is noted here that with 50 GHz channel spacing the baud rate can reach 112 GBd with Nyquist or raise cosine pulse shaping. Thus for PAM4 or PAM8 the aggregate bit rate can reach 200 Gbps or 300 Gbps with a shorter distance at which the OSNR can be satisfied. Such a short distance and ultra-high rate can be most useful for intra-C networking or the client side, i.e. access super-nodes.

We note here also that such a comb source can be employed in testing the spectra of multi-channel systems such as EDFA operating in the C+ spectral window with a total time span of less than 1.0 ms, but this is not within the scope of this chapter/book. In this chapter the comb generation source, described in detail, is a re-circulating type. There are many types of comb generators and the number of comb lines can vary from four to several hundreds. This chapter focuses on the application for transmission of ultra-high capacity and naturally the comb lines are in the C^{+}-band of the 1550 nm spectral window.

4.1 Introduction

Superchannels reaching a total capacity of a few to tens of Tbps have attracted significant attention for the emerging optical transport networks. Comb generation of narrow line optical sources of high stability and minimum jitter from a single primary source offers significant advantages in superchannel optical transmitters [1–3]. This section presents the generation and modulation of superchannels defined as multi-channels generated by the lightwaves coming from the same reference laser source or the primary laser line, by generation of multi-carrier laser (MCL) sources from primary lightwave carrier re-circulating sideband (SSB) frequency shifting techniques. The uniqueness of the SSB generation presented here is the synchronous modulation of the lightwave field re-circulating around the loop which incorporates an integrated optical synchronous modulator (IOSM) composed of a master–slave interferometric intensity or phase modulation. Their electrodes are controlled in such a way that the synchronization of the modulated lightwaves can be in constructive interference, hence maximizing the power of the generated sub-carriers. The formation of Tbps superchannels (SCs) can be 10×100 G (1 Tbps) or 20×100 G (2 Tbps) with the amplitudes of the sub-channels equalized using wavelength selective switches. In order to achieve efficient spectral packing of the modulated channels, pulse shaping of the data sequence is also implemented using a DAC at a sampling rate of 56–64 GSa s^{-1}, so that the modulated spectra of the sub-channels exhibit 'brick-wall like' shapes, hence minimizing the channel crosstalk. By employing the sub-carriers of the MSCL as laser sources, modulation of individual sub-carriers is implemented, with and without pulse shaping, to generate SCs in which each individual channel can be modulated using polarization multiplexing techniques and the QPSK modulation format. Thus the two polarized channels at 28 GBd and QPSK (2 bits/symbol) gives 112 Gbits s^{-1} (Gbps). So if 10 or 20 sub-carriers of the comb generator are employed, then the resulting aggregate bit rate reaches 1.0 Tbps and 2.0 Tbps. These superchannels can be generated for transmission over a long haul multi-span optically amplified fiber line which consists of several 80 km spans of non-dispersion compensating standard single-mode optical fibers (SSMFs).

So if combining both the spectra of the shifting left and shifting right with respect to the central original carrier frequency position, then a dual sideband shifted sub-carrier comb generation can be achieved. This type of comb generator is demonstrated using two re-circulating frequency shifting ring comb generators integrated with an optical switching and coupling sub-structure. This is possible provided that there is no interference between the previously generated and the current generation, hence allowing the generation of modulated channels whose total capacity reaches the limit and beyond of 25 Tbps. To the best of the author's knowledge, the dual band comb generator is experimentally demonstrated here for the first time.

The rest of this chapter is organized as follows: section 4.2 outlines the principles of operation of the comb generation with the frequency shifted to either the left or right side of the original lightwave carrier. The experimental results obtained for such comb generated lines are given, in particular the measured linewidth of these comb lines. Section 4.3 presents our experimental observation of the generation of dual sideband sub-carriers using a double ring configuration. Tbps superchannel optical transmitters implemented using these SSB comb generators are reported in section 4.4. Their integration into optically amplified fiber transmission systems is also described and the transmission performance of these transmission systems is then given. Finally some concluding remarks and potential future works are provided (figure 4.1).

4.2 Comb generation of multi-sub-carriers

4.2.1 General structure

The schematic of a constructed re-circulating frequency shifting (RCFS) comb generator is shown in figure 4.2, in which a re-circulating loop is the basic configuration incorporating an optical modulator, an optical amplifier to compensate the insertion losses of various inline devices and an optical attenuator whose coefficient can be adjusted so that the overall loop gain is slightly less than unity so as to avoid self-oscillation or lasing. Recently such a re-circulating comb source has been demonstrated in [4, 5], whose schematic and generated spectra are shown in figure 4.1. A 50:50 optical coupler or 3 dB split is employed for coupling into and out of the loop of the continuous lightwaves and combed sub-carrier lines, respectively. This fiber coupler is to be replaced by an optical switch and associate couplers for generation of shifted carriers to both sides of the main lightwave carrier, the dual shift re-circulating comb generator, which will be described in section 4.3. An electrical synthesizer generating sinusoidal waves whose frequency determines the frequency shifting of the optical lines of the generated comb is used to apply into the electrodes of the modulator, hence using the electro-optic effect to modulate the phase of the guided lightwaves and thus using optical modulation.

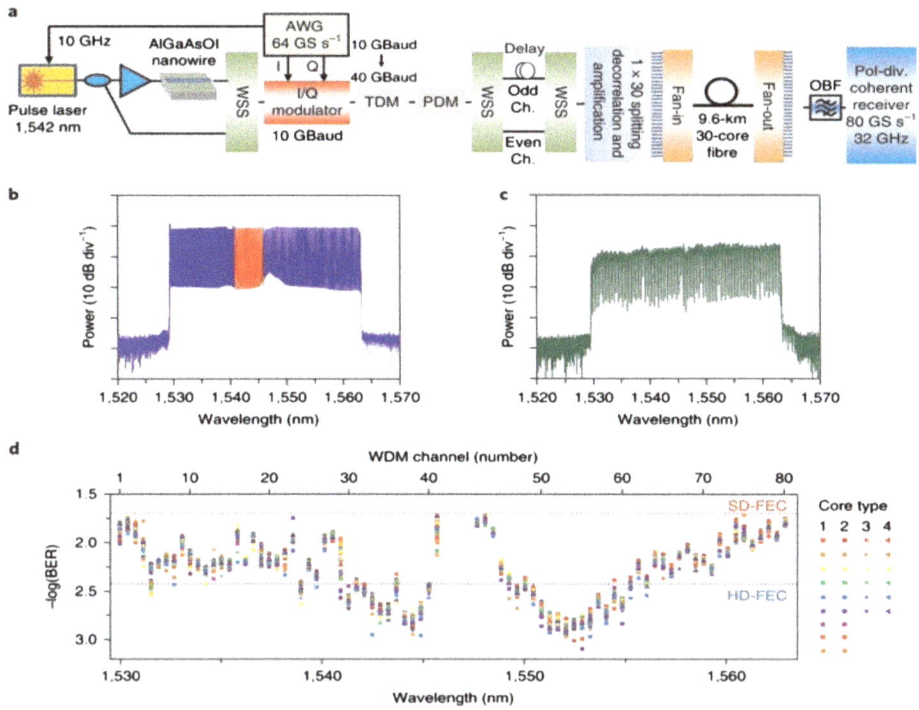

Figure 4.1. Schematic and generated spectra of a high density comb line laser: (a) comb source and system schematics; (b) channel spectra prior to transmission, or launched into the fiber; (c) channel spectra after transmission; and (d) channel transmission performance.

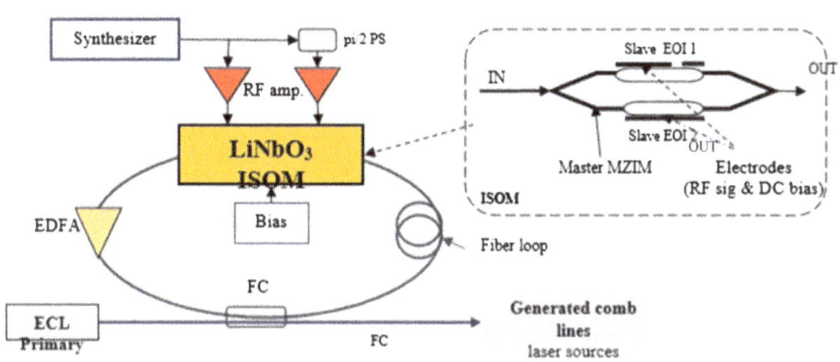

Figure 4.2. Schematic structure of the RCFS comb generator. ECL: external cavity laser; EDFA: Er-doped fiber amplifier; FC = fiber coupler; PS = phase shifter; ISOM = interferometric synchronous optical modulator.

4.2.2 Synchronous optical modulator

The interferometric synchronous optical modulator (IOSM) incorporated in the loop is a synchronous type whose structure can be seen in the inset of figure 4.2, and is composed of a master Mach–Zehnder interferometric modulator (MZIM) whose two arms are connected by two slave electro-optic interferometers (EOIs). The lightwave going through the two slave interferometers can be biased and activated such that constructive rather than destructive interference can be achieved at the combined output. The optical waveguides of the ISOM are formed by diffusion of thin Ti film deposited on top of the Z-cut or X-cut $LiNbO_3$ substrate. Optical waveguides support only single-mode and both TE and TM polarization. However, only one polarized mode is excited in the circulating ring. A thin SiO_2 film is deposited on top prior to the deposition of electrode film to match the velocity of the RF traveling wave to that of the guided optical waves. Traveling wave electrodes are independent with respect to the DC electrodes (i.e. unconnected) so that independent biasing to create the initial phase states of the guided optical waves can be achieved. In addition to the two DC electrodes deposited in one light path of the slave EOI, another dc electrode is available in one path of the master MZIM and ideally at the output of one of the slave EOIs so that the tuning of the phase of the interfered lightwaves emerging from the outputs of the two slave EOIs can be achieved in synchronization. Thus this modulator is termed as the interferometric synchronous optical modulator (ISOM).

As depicted in the inset of figure 4.2, the two arms of IOSM are driven with a $\pi/2$ phase shift with respect to each other so that single sideband (SSB) shifting of the lightwave can be achieved at the output of the modulator. Implementing the $\pi/2$ phase shift is indeed performing a Hilbert transformation. The principles of the shifting of the optical carrier frequency to the right or to the left of the original lightwave carrier are shown in figure 4.3. Applying sinusoidal or co-sinusoidal RF signals to the electrodes of the slave EOIs will shift the lightwave carrier to the right or left, respectively, as the summation of the two modulated lightwaves emerging at the output of the slave EOI are constructively or destructively combined in either

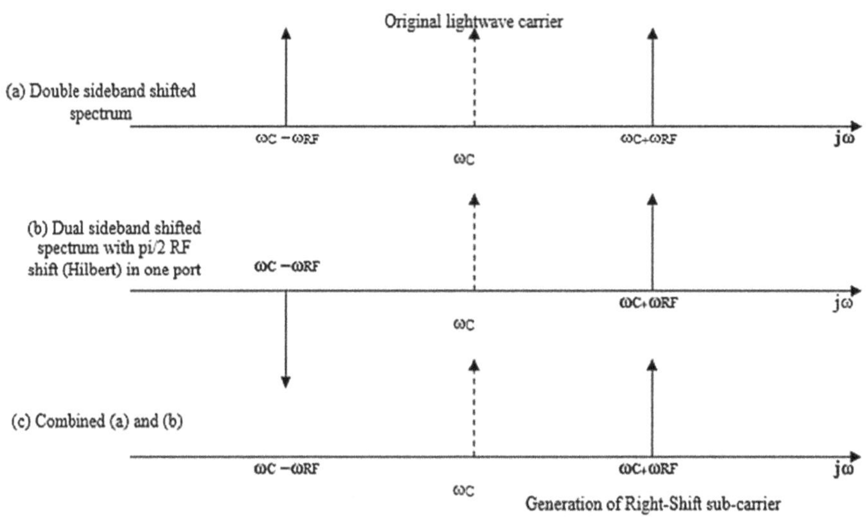

Figure 4.3. Spectra of (a) the RF modulated carrier, (b) the $\pi/2$ phase shift or Hilbert transformed RF wave driven signal modulated lightwave and (c) the constructive and synchronous spectrum, i.e. the generated right shift sub-carrier.

band of the spectrum, thus resulting in single sideband (SSB) frequency shifting. The SSB shift is critical so that multiple lines or interference of the other sideband sub-carriers with the shifted frequency sub-carriers generated in the precedent modulation can be avoided. It is noted here that the SSB principle was first demonstrated by Smith *et al* [6] to transport RF signals over a long length fiber. In our re-circulating comb generator, the optical wave is shifted and then circulating through the loop and then shifted again and again to generate other multiple shifted sub-carriers.

Thus the SSB modulation serves two purposes: first to shift the original carrier to only the right or the left side of the original carrier, and second to minimize the interference of the other sideband generated carriers if double sideband (DSB) modulation is employed. This SSB modulation technique was also employed by several research groups for transmission over long-distance fiber to carry radio channels [7–9] or digital channels using orthogonal frequency division multiplexing (OFDM) [10–12] and in the comb generation for Tbps transmission, but not in synchronization [13–16].

Therefore, for the case of dual band shifting of the lightwave main carrier, described later in section 4.3, it would require two comb generators combined with an optical switch and optical couplers to generate the shift left and shift right sub-carriers with respect to the central original lightwave, following the principles illustrated in figure 4.3(b), (c), with both the normal and the reverse $\pi/2$ phase shift.

4.2.3 Implementation of the comb generator

The comb generator is implemented using fiber based devices. A ring of fiber of length about 25 meters is formed with one input and one output port of a 3 dB optical fiber coupler (FC). This means that there is always a 50% intensity of an RF shifted lightwave coupled back into the ring and 50% coupled to the output port. A $LiNbO_3$ ISOM is incorporated in the fiber ring. The ring length is calculated including the length of the fiber-pig-tailing sections of all optical components associated within the ring. The overall gain of the ring must be less than unity so as to avoid any lasing. Er-doped fiber amplifiers are included so as to compensate for the optical loss of the fiber, the modulator and the coupler FC. The open loop is constructed first and the modulation of the lightwave carrier by a purely sinusoidal wave from a reference synthesizer is applied. The shifting of the lightwave carrier to the left or to the right is observed, as shown in figure 4.4(a), (b), respectively. Note that this frequency shifting can be either shifting left or right depending on the relative phase shift applied to the two slave EOIs embedded in the two arms of the master MZIM. As observed, the suppression of the residual lightwaves can reach higher than 25 dB by the tuning of the biasing voltage applied to the arms of the EOI, in particular the dc bias of the master MZIM. This is why synchronization of the optical modulator is essential for an efficient and high carrier-to-noise ratio (CNR) to be achieved. Under the synchronization scenario a minimum phase noise can be obtained. This is very important for these comb lines to be employed as low noise sources for long distance and coherent reception of the modulated sub-carriers over many thousand kilometers of dispersion

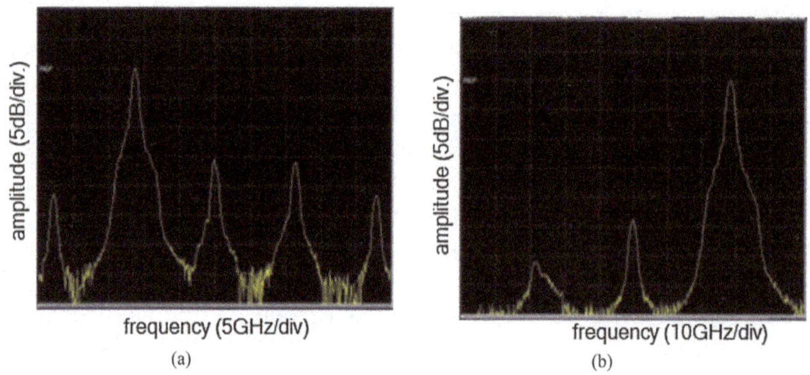

Figure 4.4. Spectrum of the SSB lightwave at the output of the synchronous optical modulator: (a) shifting right and (b) shifting left in the frequency scale. The resolution of the horizontal scale is 40 pm/div. for (a) and 80 pm/div. for (b), and 5 dB/div. in the vertical scale for power.

non-compensated fiber. These residual sub-carriers must be at this level so that noises due to the interference with other generated sub-carriers can be considered negligible.

The control of the stability of the generated comb lines is implemented by a control feedback mechanism with feedback signals coming from the synchronization of the jittering of the referenced RF synthesizer, the injection optical power level into the re-circulating ring, and the pumping level of the pump laser into the optical amplifying devices incorporated in the ring. To set the spectral spacing between the comb lines, a synthesizer launches the RF waves to the two arms of the EOIs with RF amplifiers to boost the electrical signals to an appropriate level so that the maximum modulation depth can be achieved with minimum nonlinear effects. We note that although the modulator 3 dB bandwidth is specified at 23 GHz, it is still possible to operate the modulator at 33 GHz, although with some RF signal loss, hence demanding higher RF signal power. This is because the driving signal is purely sinusoidal and has a very narrow band. We have successfully generated comb lines from 28 GHz to 33 GHz for channel spacing between sub-carriers of 28 G, 30 G and 33 Gb s^{-1}, as shown in figure 4.5. Note also in figure 4.5(b) the residual original carrier in the far right of the spectrum. This appears because of the use of moderate gain optical amplifiers (specifically EDFAs) and it takes some time for the optical signals to reach a saturation state in the amplifier. Depending on the loop and the gain of the EDFA with respect to those components embedded in the re-circulating ring, there may be one or two residual frequency shifted lines in the spectrum, as is observable in figure 4.4. The spectrum is flat over some considerable spectral range, about 5.5 nm, which is sufficient to support ten sub-carriers whose equal spacing is of the order of 50 GHz (or equivalently 0.4 nm) in the 1550 nm region. With these ten sub-carriers and the symbol rate of 28 GBds s^{-1} per sub-carrier channel, the aggregate bit rate is 100 Gbps per sub-carrier if using two polarized channels and QPSK or 2 bits per symbol. Therefore a total capacity of 1 Tbps can be obtained for the ten sub-carriers generated by this comb generator. For 2.0 Tbps we must design the conditions for the comb generator such that 20 sub-carriers can be generated.

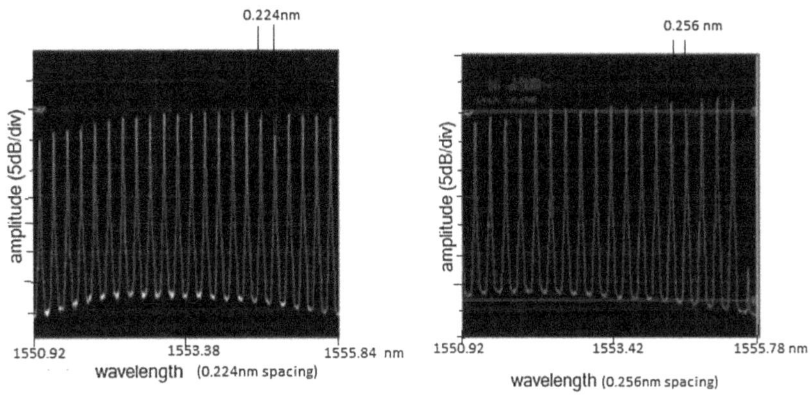

(a) (b)

Figure 4.5. Spectrum of the multi-sub-carrier comb lines generated by the RCFS loop. Vertical scale: power amplitude of arbitrary units; horizontal scale: frequency of 50 GHz/div.

The comb lines can also be equalized using a wavelength selective switch (WSS) to obtain amplitude equalized sub-carriers, the ripple level of the comb lines depends on the flatness of the optical amplifier incorporated in the ring, thus a flat spectral EDFA would normally be selected for this type of re-circulating RF shifted comb generator. On both the left and right shift spectra, the CNR is achieved in the order of 27 dB. This is quite remarkable compared to the other comb generated lines reported so far [17, 18]. We believe that this is contributed by the synchronization of the modulated lightwaves passing through the master MZIM and slave EOIs.

A low phase noise external cavity laser (ECL) is used as the primary lightwave which is fed into the SSB frequency shifted fiber ring under non-resonant conditions to generate comb sub-channels. By non-resonance we mean that the resonance condition for lasing is not permitted by ensuring that the total gain and loss of the lightwaves circulating in the ring is less than unity. The ECL output power is set at its highest level of 13 dBm. A higher power level can be used if an EDFA booster amplifier is employed.

The spectra of the comb lines generated from the re-circulating comb generator are shown in figure 4.5(a), (b). Thus we note the power level of the SBS line at −10 dBm and −4 dBm for the comb lines, which have also been passing through an EDFA at a saturation power of 16 dBm which is distributed to all sub-carrier lines.

Both sides of the primary laser lines can also be generated by employing a dual ring structure with the SSB shift to the right or the left of the primary line as described in the next section. This ultra-wide spectral comb line source has also been generated in our laboratory.

The optical amplifier incorporated in the re-circulating ring is operating in the saturation mode only when the power level of the lines is sufficiently high to boost it into the saturated region, otherwise it would operate in the linear region. For this reason, we observe that the frequency shifting lines would not be in the saturated level for the first few sub-carrier lines, the left-most side of the spectrum (figure 4.5(a)). The tuning of the DC bias supply voltage to the electrodes is quite sensitive and would be in the tens of millivolts range. One important point we must draw attention to is that the output fiber port must be of the angled connector type so that no reflection of sub-carrier combed lines can feed back into the re-circulating loop to avoid the noise oscillation of the generated comb lines. Note that the suppression of the other harmonic lines is more than 25 dB.

The comb generator described in this section is then employed in a multi-Tbps transmission system which will be described in section 4.4.

4.3 Dual band frequency shifting re-circulating comb generator

Further to the possibility of generation of comb sub-carriers over the whole C-band in the region 1520 nm to 1565 nm for optical communication systems with the referenced central lightwave carrier located at the center of the band, the frequency shifting of the original carrier must be into both the left and right sides of the carrier. We thus propose a dual band comb generator structure in which two re-circulating frequency shifting ring comb generators share 'a common optical switching structure', as shown in figure 4.6.

Further detailed principles of the experimental generation of the dual band comb lines can be described as follows. First the continuous laser line is fed into a 2 × 2 fiber coupler (FC) (one port left unused) whose outputs can then be fed to 2 × 2 optical switches, as shown in figure 4.6. The integrated optical guided-wave switch is shown as an inset of this figure. The optical switch is of $LiNbO_3$ type and fabricated by the Ti in-diffusion technique [19, 20]. The electrical switching signals are fed into the electrical port of the OSWs, as shown by a dotted line in the figure. All optical paths are shown as continuous lines. The bandwidth of the OSW is 18 GHz, which allows the switching of 100 ns to be achieved without any difficulty. The switching electrical signals are complementary to each other, indicating that when one switch

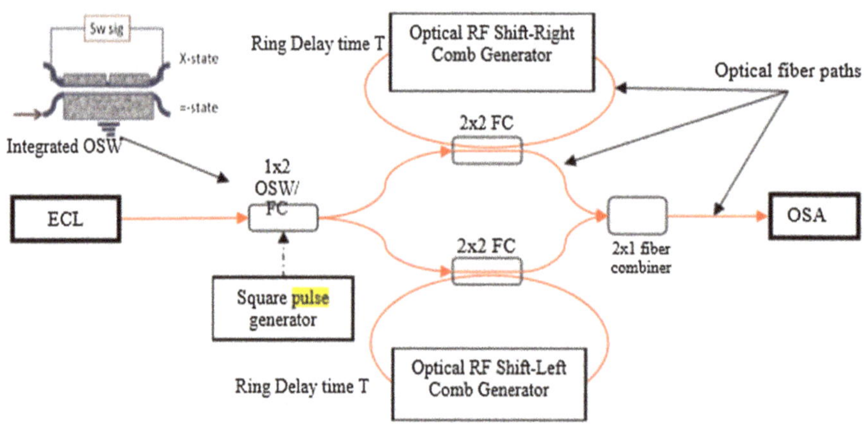

Figure 4.6. Schematic of the experimental dual band comb generation. OSW = optical switching device with a switch time much shorter than that of the ring propagation delay time *T*; FC = fiber coupler; OSA = optical spectrum analyzer. All continuous lines are optical paths. Dotted lines are electrical lines. The inset shows the structure of an integrated OSW.

is on the other is off. The switching period of the electrical switching signal is just less than that of the circulating delay time which is estimated to be 100 ns for the 20 m long fiber loop of each comb generator. Each output of the 1×2 OSW is fed into a comb generator whose RF signals are fed into an ISOM with the $\pi/2$ shift in reverse with each other in the RF signals applied to electrical ports so that the shift left and shift right of the sidebands can be obtained as described in section 2.2.

The lightwaves are injected at a power level such that the optical amplifiers incorporated in the ring, similarly to that given in figure 4.2, can operate in the saturation mode. The outputs of the two ring comb generators are then combined via another passive FC. The output of the dual band comb generator is monitored by an optical spectrum analyzer (OSA). The switching time of the optical switch must be much less than that of the ring delay time so that no interference will occur. It is obvious that as the delay time of the lightwaves circulating through a loop is sufficiently long, about 75 ns for a 25 m length ring. Thus a moderate bandwidth optical switch can be employed. Such electro-optic switches of a 3 dB bandwidth of 20 to 30 GHz in $LiNbO_3$ are commercially available. These switches can thus provide the switching time between shifting left and right within 50 ps to 33 ps[2] [21, 22].

The observed dual band spectral comb lines are shown in figure 4.6. We note here that as the switching efficiency of the OSW is not 100% but about 95%, there are some residual optical carriers which are embedded in the shifted optical sub-carriers which could cause some interference noise in subsequent comb lines. Thus we expect some degradation to the performance BER versus OSNR compared to those depicted in figure 4.17 (described later in section 4.4). We also observed some undulation of the amplitudes of the comb lines shifted to the left side due to unequal transmission loss in the re-circulating loop. Indeed, we have also set up the dual ring comb generator using a passive fiber coupler (FC) instead of the OSW shown in figure 4.6. The optical spectrum of the comb sub-carriers can also be observed in figure 4.7. We note here that the optical amplifiers embedded in both comb rings of the dual ring comb generator are implemented with a high output power of about 20 dBm and they are set such that they always operate in the saturated region under the optical launched power. Hence the comb lines reach the saturated power level in the first circulation. The difference between the switched configuration using OSW and the passive FC is that some residual lines at switching speed would appear in the spectrum when it is monitored by a high resolution spectrum analyzer. However, under the spectrum monitored by the OSA of minimum resolution of 0.1 nm these residual lines cannot be seen. The continuous original laser lightwaves are fed into the two rings alternatingly so that the first-order shifting of the optical sub-carrier can be generated at all times and hence the higher order shifted lines are generated due to the re-circulating and frequency shifting via the ISOM. The two ISOMs are modulated with the $\pi/2$ phase shift in the opposite way to that described in section 4.2. In figure 4.7 the shifting left and shifting right are referred to in

[2] See the details for 18 GHz optical switches at http://eospace.com/pdf/EOSPACE-custom-optical-switch.pdf.

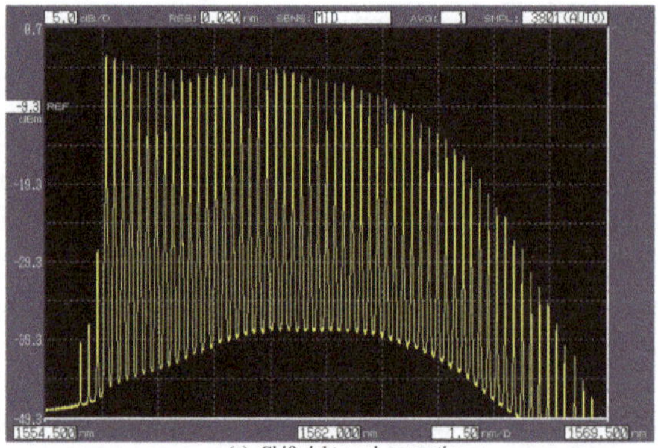

(a) Shift right comb generation

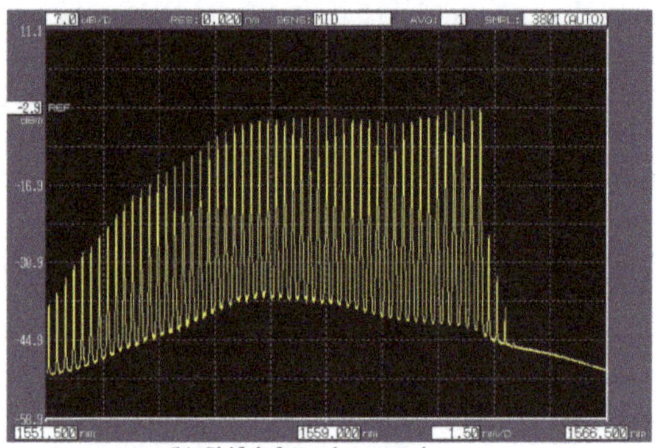

(b) Shift left comb generation

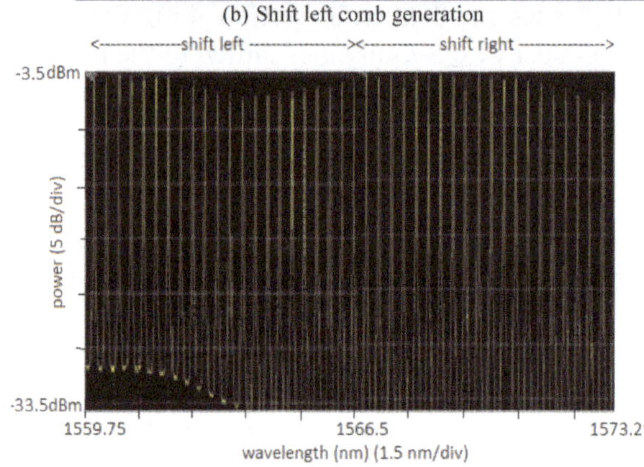

(c) Combined left and right shift spectrum of comb generation, part of the complete spectrum.

Figure 4.7. Observed optical spectra of a dual ring comb generator with double sideband shifting, sub-carriers are shown as combined RF frequency shifted left and shifted right sub-carriers: (a) right shift; (b) left shift; and (c) combined right and left comb generation, a section of the spectrum only.

frequency terms, so if they are specified in terms of wavelength then the remarks must be in reverse.

A transmission experiment is yet to be conducted for these dual band comb sub-carriers.

4.4 The comb generator in a multi-Tbps optical transmission system

The comb generator can be employed to generate the referenced laser comb lines which are then modulated with Nyquist pulse shaping implemented in a Fujitsu 56 GSa s^{-1} DAC [23]. The principal purposes of Nyquist shaping are to achieve a more effective bandwidth and minimize the inter-symbol interference (ISI) between adjacent symbols (figure 4.8). The modulated spectra by Nyquist shaping and non-shaping obtained in the experiment are shown later in this section, see figures 4.9(a), (b). However, it is well known that a direct relationship between a rectangular spectrum and the sin x/x impulse response indicates that for a sequence of Nyquist pulses, the zero crossing points occur at the sampling point. In practice it is not that simple to generate a Nyquist impulse or pulse sequence due to the nature of the sharp rectangular shape of the spectrum, but their approximately equivalent cosine functions can substitute [24]. Furthermore, the bandwidth of the DAC and that of the ISOM is not infinitely wide, so it is more realistic that the raised cosine pulse shape is employed.

Experimentally, the comb generator described above with 10 to 20 sub-carriers can be used for creating modulated superchannels of an aggregate bit rate of 1 and 2 Tbps, respectively, with spacing as closed as possible. We have demonstrated these Tbps superchannel optical transmitters in transmission systems over a 1500 km or 3500 km multi-span optically amplified non-compensating line consisting of cascade spans of standard single-mode fibers (SSMF) incorporating optical amplifiers and no dispersion compensating fibers (DCF), depending on the modulation formats of either quadrature phase shift keying (QPSK) or quadrature amplitude modulation of 16 points square constellation points (16QAM). The pulse shaping is essential, implemented as described in the next section, for close packing the channels so that more information can be transmitted over a limited spectral window. These

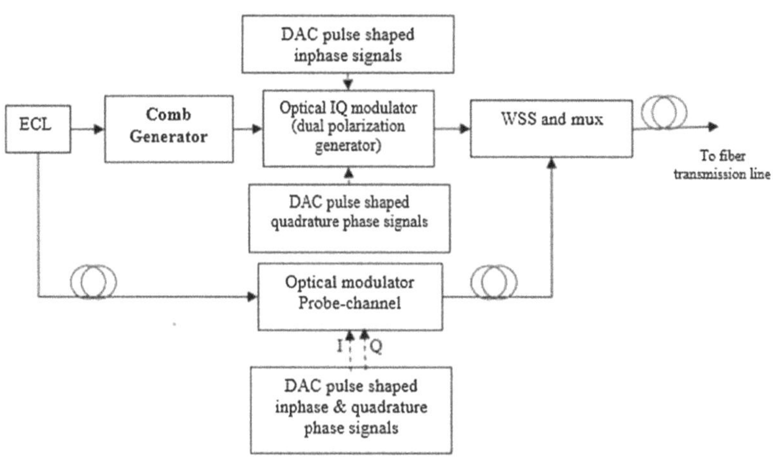

Figure 4.8. Optical transmitter employing comb generated sources and pulse shaping modulation. DAC = digital to analog converter; ECL = external cavity laser; WSS = wavelength selective switch; mux = multiplexer; continuous line = optical path; dotted lines = electrical paths.

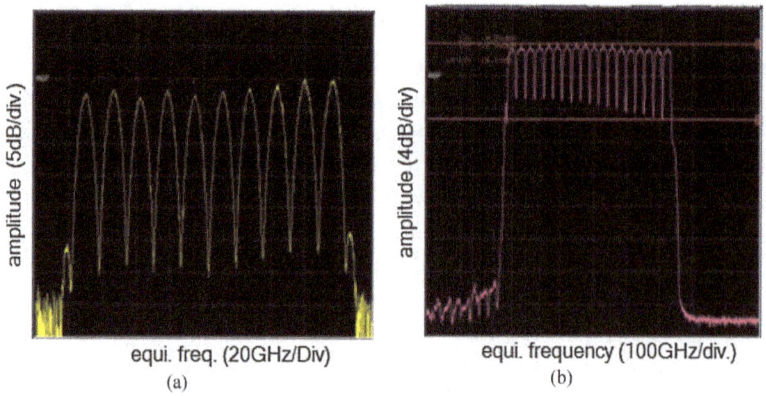

Figure 4.9. Spectrum of QPSK superchannel modulation with narrow spacing. Eighteen channels under Nyquist pulse shaping after transmission over 2000 km optically amplified SSMF spans: (a) non-shaping QPSK modulation ten sub-channels of 28 GS s^{-1} polarization multiplexed QPSK for 1 Tbps superchannels and (b) Nyquist shaping superchannels, 20 at 28 Gb s^{-1} for 2 Tbps.

superchannels are now attracting lots of interest in flexible grid and bandwidth optical transport networks [25, 26].

4.5 Packing modulated comb channels in the frequency domain using Nyquist shaping

4.5.1 Analytical representation for pulse shaping

In ultra-high capacity optical transmission systems, the higher the compact frequency spectrum of the signals is, the better the spectral efficiency one can achieve. As the time domain pulse relates well to its spectrum, the shaping of the pulse sequence can be implemented using a high speed sampling DAC and DSP. A rectangular spectrum is the ideal case as it is 'brick-wall-like', so if such an ideal spectrum of the system response is desired then its corresponding equivalent impulse response must take the well-known shape of a sinc(t) function [27]. At the sampling instants $t = kT(k = 1, 2....N)$, the amplitudes reach zero. This implies that at the ideal sampling instants, the inter-symbol interference (ISI) from neighboring symbols is negligible, or free of ISI. It is well known that a Nyquist pulse corresponds to a brick-wall-like frequency spectrum; hence one can pack channels whose spectra are close to each other and their zero crossing of the time domain response passes through one another. Note that the maximum of the next pulse increase is the minimum of the previous impulse of the consecutive Nyquist channel.

Considering a one sub-channel carrier 25 GBd PDM-DQPSK signal, with the aggregate capacity of 100 Gbps per sub-channel with two multiplexed polarizations, to reach 1 Tbps ten sub-channels would be required. To increase the spectral efficiency, the bandwidth of these ten sub-channels must be densely packed together. The most likely technique for packing the channels as close as possible in the frequency with minimum ISI is Nyquist pulse shaping, which was described earlier in this section. However, in practice such a brick-wall-like spectrum cannot be realized, thus an equivalent and very closely approximated function can be employed, the raised cosine shape [15]. Mathematically speaking, the raised cosine filter can be implemented using a low-pass Nyquist filter whose vestigial symmetry property can be the main feature, indicating that its spectrum exhibits odd symmetry about $1/2T_s$, where T_s is the symbol-period. Its frequency domain representation is a brick-wall-like function. This frequency response is characterized by two values: β, the roll-off factor, and T_s, the reciprocal of the symbol rate in Sym s^{-1}, that is $1/2T_s$, the half bandwidth of the filter. The impulse response of such a filter can be obtained by analytically taking the inverse Fourier transformation of equation (4.1) in terms of the normalized sinc function as

$$H(f) = \begin{cases} T_s & |f| \leqslant \frac{1-\beta}{2T_s} \\ \frac{T_s}{2}\left[1 + \cos\left(\frac{\pi T_s}{\beta}\left\{|f| - \frac{1-\beta}{2T_s}\right\}\right)\right] & \frac{1-\beta}{2T_s} < |f| \leqslant \frac{1+\beta}{2T_s} \\ 0 & \text{otherwise} \end{cases} \longleftrightarrow h(t) \tag{4.1}$$

$$= \sin c\left(\frac{t}{T_s}\right)\frac{\cos\left(\frac{\pi\beta t}{T_s}\right)}{1 - \left(2\frac{\pi\beta t}{T_s}\right)^2}$$

with $0 \leqslant \beta \leqslant 1$,

where the roll-off factor, β, is a measure of the excess bandwidth of the filter, i.e. the bandwidth occupied beyond the Nyquist bandwidth as from the amplitude at $1/2T$. The spectrum of a raised cosine pulse with various roll-off factors is well known. In our transmission systems the pulse shaping is conducted with a roll-off factor less than 0.3.

For a sequence of symbols, the Nyquist filter offers the property of eliminating ISI, as its impulse response passing through the zero crossing points at all nT (where n is an integer), except when $n = 0$. Therefore, if the transmitted waveform is correctly sampled at the receiver, the original symbol values can be recovered completely. However, in many practical communications systems, a matched filter must be used at the receiver, so as to minimize the effects of noise. For zero ISI, the net response of the product of the transmitting and receiving filters must equate to $H(f)$, thus we can write

$$H_R(f)H_T(f) = H(f), \tag{4.2}$$

where $H_R(f)$ and $H_T(f)$ are the transfer functions of the receiving and transmitting filters, respectively.

Alternatively we can rewrite equation (4.2) as

$$|H_R(f)| = |H_T(f)| = \sqrt{|H(f)|}. \tag{4.3}$$

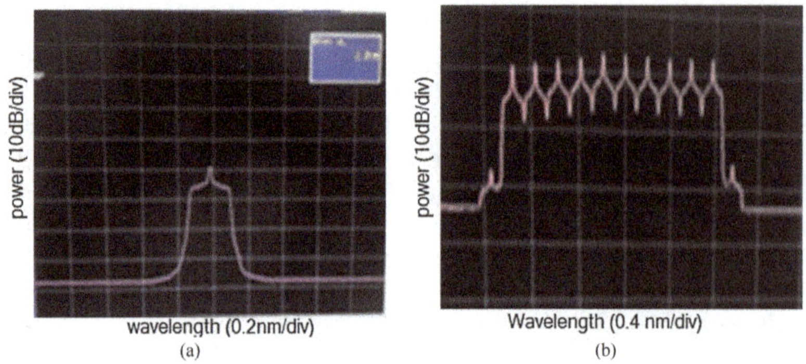

Figure 4.10. Spectrum of (a) an extracted spectrum of a modulated channel using Nyquist shaping monitored after the optical filter (see also figure 4.11) and (b) packing of ten comb generated sub-carriers. Note that at the wavelength 1550 nm and neighboring spectral region 0.8 nm is approximately equal to 100 GHz.

The filters which can satisfy the conditions of equation (4.3) are the root-raised cosine type. The main problem with root-raised cosine filters is that they occupy a larger frequency band than that of the Nyquist sinc pulse sequence. Thus for the transmission system we can split the overall raised cosine filter with one root-raised cosine filter each at both the transmitting and receiving ends, provided the system is linear. This linearity is to be specified accordingly. An optical fiber transmission system can be considered to be linear if the total average power of all channels under simultaneous transmission is under the nonlinear self-phase modulation (SPM) threshold limit [28] which is about 10 dBm for SSMF. When it is over this threshold a weakly linear approximation can be used and some nonlinear distortions will occur. Therefore the adjustment of the power level of each sub-carrier generated by the comb source is very critical to avoid this nonlinear distortion effect (figure 4.10).

The design of the Nyquist filter influences the performance of the overall transmission system [29]. The oversampling factor, roll-off factor for different modulation formats, and the design of the finite impulse response (FIR) Nyquist filter and its key parameters are determined to optimize for the case when sub-carriers of the comb generator are employed. We use a wavelength selective switch (WSS) to equalize the power level of all sub-carriers for this purpose (see figure 4.11). If taking into account the transfer functions of the overall transmission channels including the fiber, the WSS and the cascade of the transfer functions of all optical-to-electrical (O/E) components, the total channel transfer function is more Gaussian-like. To compensate this effect of the DSP based transmitter (DSP-Tx), a special Nyquist filter acting as a pre-emphasis filtering is cascaded to achieve the overall frequency response which is equivalent to that of the rectangular or raised cosine filter with a roll-off factor of 0.9.

4.5.2 Implementation

The experimental implementation of the comb generator in a multi-channel optical transmitter for Tbps transmission is shown in figure 1.10. The comb lines can be split in the optical domain and individually modulated and then combined to launch into

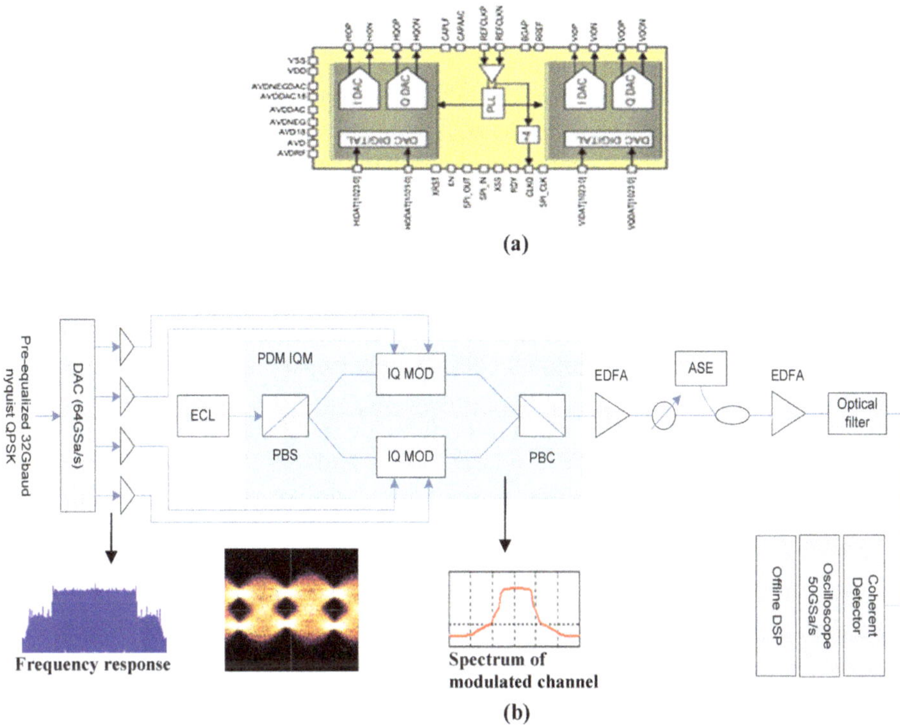

Figure 4.11. (a) Functional block diagram of an integrated 56–64 GSa s^{-1} (courtesy of Fujitsu Co. Ltd). (b) Experimental set-up of a 118 Gb s^{-1} Nyquist PDM-QPSK transmitter and back-to-back performance evaluation. ECL = external cavity laser; PBC = polarization beam combiner; PBS = polarization beam splitter; PDM IQM = polarization division multiplexed in-phase-quadrature phase modulator; MOD = modulator; EDFA = Er-doped fiber amplifier; ASE = amplification stimulated emission noise.

the optical fiber transmission line. Alternatively, as a means of proving the principle, the complete set of comb lines of sub-carriers can be modulated using one optical modulator. One channel of these modulated channels is then filtered by a WSS and then independently replaced by a probe modulated channel with respect to the combed channels. The observed Nyquist pulse shaped spectrum of the modulated comb sub-carriers are shown in figure 4.10(b) for 18 generated sub-carriers. Figure 4.10(a), (b) shows, respectively, the experimentally observed spectrum of a raised cosine modulated sub-carrier obtained in an optical spectrum analyzer (OSA) and ten modulated channels for 1.0 Tbps field trial transmission. The dual polarization multiplexed I–Q modulators used in this transmitter are Fujitsu type [30] in which two parallel I–Q modulators are formed, one each for the transverse electric (TE) and transverse magnetic (TM) polarized modes. Thus, in order to provide modulation of both polarizations, two DACs must be employed. The modulation and bias electrodes of the modulators are independent of each other.

4.6 Pulse shaping using ultra-high sampling rate DAC

This subsection gives a brief insight into the pulse shaping for packing superchannels whose carriers are generated from the SSB comb generators described in section 4.2.3. The principal sub-system of the superchannel transmission system is the DSP based optical transmitter in which the DAC plays the central role in the pulse shaping, channel equalization and pattern generation. A schematic structure of the DAC and functional blocks are shown in figure 4.11(a), (b), respectively. An external sinusoidal signal is fed into the DAC so that N times the multiplying clock source can be generated for sampling at 56–64 GSa s^{-1}. Thus the jittering and clock accuracy depends on the stability and noise of this synthesizer. A stable low phase noise signal generator Agilent model N9310A is employed to provide the referenced oscillation source for the DAC[3] injected into the port REFCLKP and REFCLKN given in figure 4.11(a). Four DAC sub-modules are integrated in one unit to provide four pairs of eight outputs of $(V_I^+, V_Q^+)(H_I^+, H_Q^+)$ and $(V_I^-, V_Q^-)(H_I^-, H_Q^-)$ which are then employed to modulate the horizontally and vertically polarized optical channels of I–Q optical modulators. Thus with a baud rate of 25 GBd and 2 bits/symbol with polarization multiplexing the combined optical channel would carry 100 Gbits s^{-1}. Hence when 20% error coding is used, 25 GBd becomes 28 GBd and hence the aggregate bit rate becomes 112 Gbits s^{-1}. The Fujitsu DAC is specifically designed for this type of I–Q modulator. As shown in figure 4.11(a), there are two DACs with output sequences dedicated to positive (P) and negative (N) logics. The functionalities of the other input and output ports are dedicated to the memory loading and necessary signals, and then the biasing voltages required for the DAC operations.

The generation of the I and Q components of the modulated lightwaves can be implemented as follows. The electrical outputs from the quad DACs are in pairs of positive and negative and complementary with each other. Thus we would be able to form two sets of four output ports from the DAC development board, that is $(V_I^+, V_Q^+)(H_I^+, H_Q^+)$ and $(V_I^-, V_Q^-)(H_I^-, H_Q^-)$. The positive superscripts indicate the normal pulse sequence while the negative superscripts indicate the complementary signals. The subscripts I and Q are used for the in-phase and quadrature phase components. Each output can be independently generated with offline uploading of pattern scripts into the memory of the DAC for execution. The arrangement of the DAC and polarization division multiplexing (PDM)-I–Q optical modulator is depicted in figure 4.11(b). Note that we require two PDM-I–Q modulators for the generation of odd and even optical channels. As Nyquist pulse shaped sequences are required, a number of pressing steps are conducted to ensure the functionalities of the DAC: (i) characterization of the DAC transfer function to see whether it can handle the signal bandwidth; if not, then (ii) pre-equalization in the RF domain by implementing a pre-emphasis on the DAC generated sequence to achieve an equalized spectrum in the optical domain, that is at the output of the PDM-I–Q modulator.

[3] Fujitsu Semiconductor Europe GmbH, '8-Bit, 55-65GS/s, 0.9V/1.8V digital-to-analog converter FMLY50FEAMEDJ8A0', issued 2011.

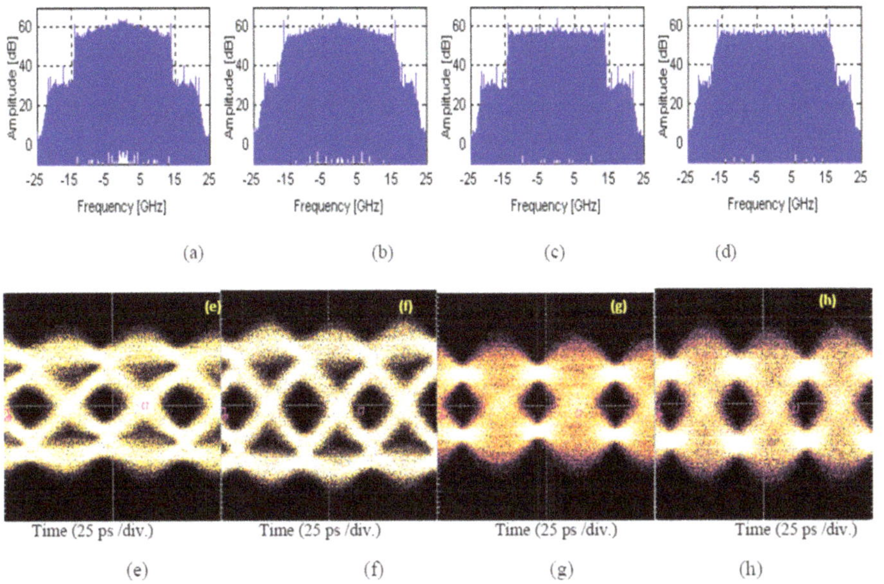

Figure 4.12. Generated spectra (a) and corresponding eye diagrams (e) of 28 GBd electrical signals using DAC without pre-equalization (b) and for 32 Gbd (f). Spectrum (c) and eye diagram (g) of 28 Gbd electrical signals after DAC with pre-equalization, (d) and for 32 Gbd (h). All observed using input to the electrical port of a sampling oscilloscope.

The characterization of the DAC is conducted by launching a sinusoidal wave to the DAC at different frequencies and measuring the waveforms at all eight output ports. As observed in the insets of figure 4.11, the electrical spectrum of the DAC is quite flat provided that pre-equalization is implemented. The spectrum of the DAC output without equalization is shown in figures 4.12(a), (b). Their corresponding eye diagrams are shown in figures 4.11(e), (f). The transfer characteristics are not flat due to the non-flatness of the transfer function of the DAC, as shown in figure 4.11, which is obtained by driving the DAC with sinusoidal waves of different frequencies and recording the output magnitudes. This shows that the DAC acts as a low-pass filter with the amplitude of its passsband gradually decreasing when the frequency is increased. This effect can originate as the number of samples per period of the sinusoidal wave is reduced when its frequency is increased because the sampling rate can only be set in the range 56–64 GSa s^{-1}. Thus a pre-emphasis is required for equalization of this DAC response.

The equalized RF spectra are depicted in figures 4.12(c), (d). The time domain waveforms corresponding to the RF spectra are shown in figures 4.12(g), (h), respectively, as observed by a real-time sampling oscilloscope Tektronix DPO 73304A or DSA 720004B. Furthermore, the noise distribution of the DAC indicates that the sideband spectra have been shifted from the noise floor by an amount which accumulates from the circulating paths through the optical amplifier in each turn. It is noted by observing the inset eye diagram of figure 4.11(b) and those given in figures 4.12(g), (h) that the zero crossings of the pulses are converging to a focal point. Thus if one can carry out the sampling of the received pulse sequence at these zero crossings then the probability of error would be minimal (figure 4.13).

4.7 Transmission of superchannels formed by modulated MCL

The optical transmitter employing the comb generator is shown in figure 4.14. The combed sub-carriers are generated with a spacing varying from 28–32 GHz to accommodate 28–32 Gbps QPSK modulation formats, respectively. The receiver is a coherent type, thus a local oscillator of the same frequency as the probe channel sub-carrier must be employed to mix with the incoming probe channel to bring the data sequence back to the baseband. The roll-off factor of the raised cosine filter for all sub-channels is set at 0.9. The optical spectrum is very close to the raised cosine shape. Due to the resolution of the OSA the exact roll-off factor cannot be determined accurately but only a close approximate. Figure 4.15 shows four sections of the DSP based coherent receiver. First, the superchannels are fed into a tuneable optical filter through which a particular channel is selected for the measurement of its transmission performance. The filtered channel then goes through a 90° optical hybrid coupler in which the two polarized channels are separated. Furthermore, via the use of a $\pi/2$ optical phase tuning both the in-phase and quadrature phase

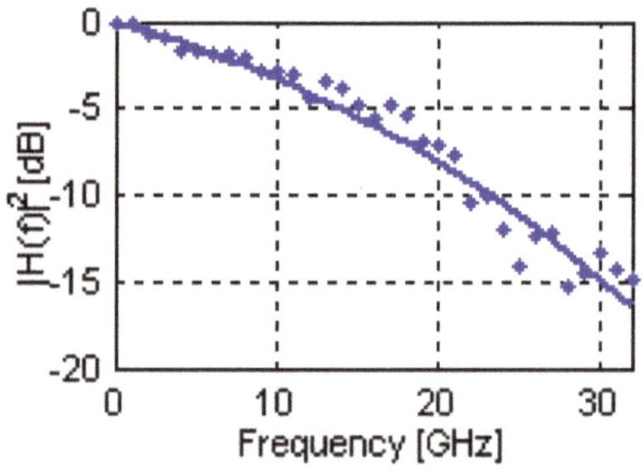

Figure 4.13. Frequency transfer characteristics of the DAC. Note the near linear variation of the magnitude as a function of the frequency.

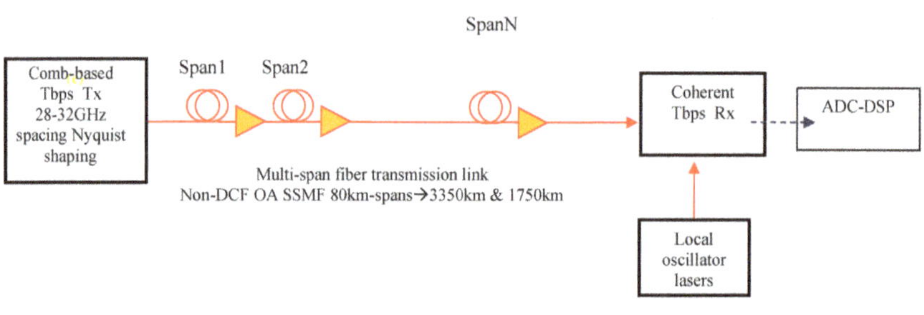

Figure 4.14. Optical transmission system employing comb generated sources and coherent reception techniques in association with digital signal processing. DAC = digital to analog converter; ECL = external cavity laser; Tx = transmitter; dotted line = electrical line; continuous line = optical path.

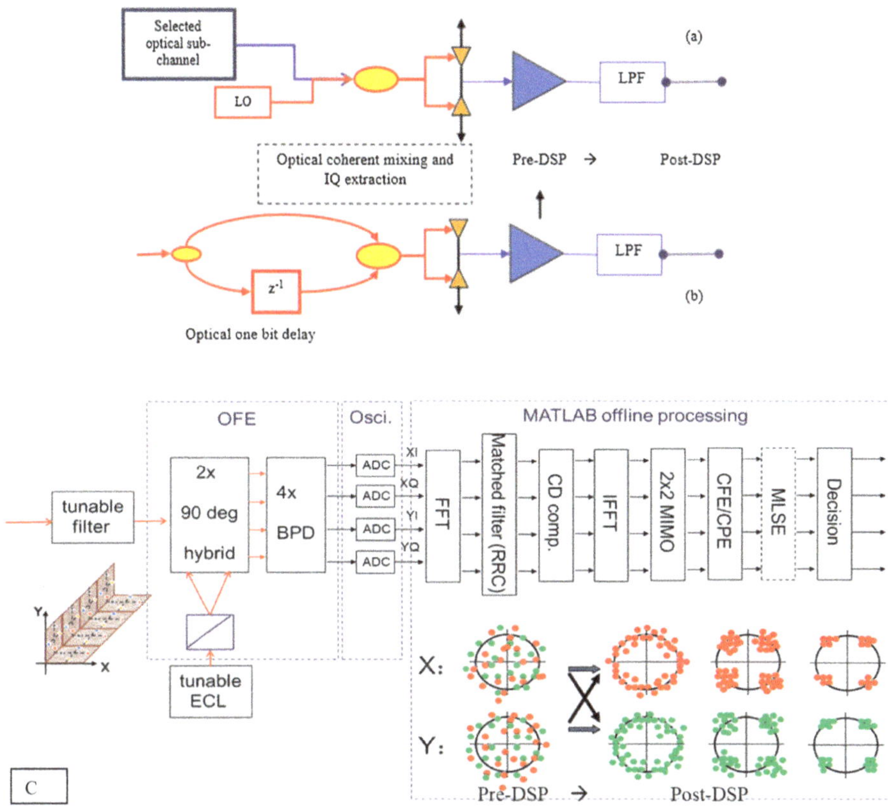

Figure 4.15. (a) Generic schematic of the coherent receiver. (b) Self-coherent receiving system. (c) Details of coherent reception and digital signal processing. ADC = analog to digital convertor; MIMO = multi-input multi-output; IFFT = inverse fast Fourier transform; CFE = carrier phase estimation; MLSE = maximum likelihood sequence estimator; CD = chromatic dispersion; BPD = photodetector pairs and electronic pre-amplifier; ECL = external cavity laser.

components are also extracted and then coherently detected by mixing with a local oscillator laser.

At the coherent receiver, a comb generator can be employed to mix with all sub-channels. However, due to the limited bandwidth of the optical detector where the mixing and beating between the local oscillator and the channel may occur, our current receiver is demonstrated by demultiplexing all sub-channels and, one by one, the probe channel is recovered in the electronic domain for processing. The outputs of the hybrid coupler are detected by a pair of photodetectors connected back-to-back to maximize the openings of the eye diagram of the sequence. These electronic signals are then pre-amplified by a trans-impedance amplifier (TIA) balanced receiver. In our coherent receiver the local oscillator is tuned to the frequency of one of the sub-carriers for which homodyne reception can be achieved. However, in practice there is a mismatching in frequency of the sub-carrier of the modulated channel and that of the local laser source. This mismatching can be compensated in the digital domain using an appropriate algorithm, for example the constant modulus amplitude type [31]. Thus this type of coherent reception is commonly termed as serrodyne detection.

The electronic signals at the outputs of the balanced receivers are then sampled by a 64 GSa s^{-1} analog to digital converter (ADC) to give the received sequences in the digital domain for further processing by the DSP. The ADC sampling rate can be performed at 56–64 GSa s^{-1} for 28–32 GBd, respectively. Thus we obtain two samples per symbol/baud period that satisfy the Nyquist sampling criteria. These digital data are then processed offline by a number of DSPs in which the clock information is recovered and the bit error rate (BER) can be determined. The third block on the far right-hand side of figure 4.15 depicts the schematic of the DSP section in which individual blocks indicate the main principles of processing. First the received samples are transformed into the frequency domain by the fast Fourier transform (FFT) block, then the compensation of the distortion due to chromatic dispersion is performed and then carrier phase recovery. The samples are then processed in the discrete time domain using certain algorithms, such as the maximum likelihood sequence estimation (MSLE) or decision feedback estimation (DFE), to recover the phase constellation, as shown by the inset in the post-DSP section of figure 4.15. Finally the BER can be obtained. The cross lines between the X (horizontal polarized channel) and Y (vertical polarized channel) indicate mutual cross-processing between these two polarized channels to maximizing the recovery of the transmitted pulse sequence.

The equalization in the power amplitude of the superchannels is implemented using a wavelength selective switch (WSS). Individual comb modulated channels are coherently detected via the use of a $\pi/2$ polarized hybrid coupler and then digital processing of lines with an achieved BER in the region of 1e-3, which is acceptable for forward error-coded transmission systems. The power sensitivity is also obtained with launched power in the normal acceptable range so that no significant nonlinear impairments occur. This transmission has also been used for field trials in Deutsche Telecom and Vodafone installed fiber transmission lines located in Germany, as shown in figure 4.15.

The optical transmission path of the two field trials using the optical transmission systems employing superchannels generated by modulating the sub-carriers of the comb generator areas follows. Path 1: from Darmstadt to Nurnberg with an additional several hundred kilometers of standard single-mode fibers inserted in the laboratories at Stuttgart and Nurnberg. Path 2: from Esborn to Regensburg with additional kilometers of fibers inserted in different locations in the link. The numbers of sub-carriers generated from the comb generator employed in these field trials are 10 and 20 sub-carriers, thus 1.0 Tbps and 2.0 Tbps superchannels can be transmitted, respectively. The probing and processing of individual channels and processing of the received signals show that BER of 1e-3 or better can be achieved without any difficulty. A typical plot of the BER versus the optical signal-to-noise ratio (OSNR) is depicted in figure 4.17. It is thus proven that the comb generator is very useful in multi-Tbps optical transmission systems. In the following we describe the transmission systems in more detail.

It is noted that the optical transmission lines consist of standard single-mode optical fibers. No dispersion compensation of the fiber chromatic dispersion in the optical domain is used. All compensations or equalization of distortions are implemented in the digital domain, as shown in figure 4.15. Each of the spans of the fiber transmission line is established at 80 km, so the loss is about 19 dB over one whole span and the gain of the optical amplifiers is set at this level (figure 4.16). The optical amplifier noise value is about 5 dB. The field transmission distance was demonstrated at 3350 km and 1730 km for a 100 G and 200 G bit rate per sub-channel, respectively. For 200 G the modulation scheme was 16QAM (4 bits/symbol).

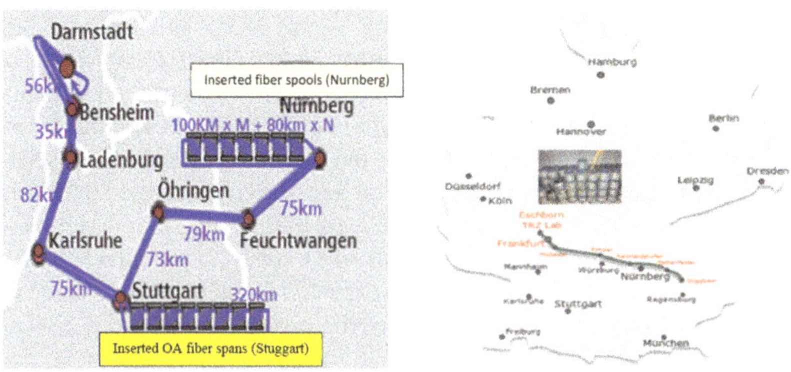

Figure 4.16. Field trial optical link path for a comb-incorporated optical transmitter and coherent reception systems: (a) Darmstadt to Nurnberg with an additional several hundred kilometers in the laboratory inserted in Stuttgart and Nurnberg. (b) From Esborn to Regensburg with an additional several hundred kilometers of fibers inserted at different locations to make up the total length of more than 3000 km, as shown along the link.

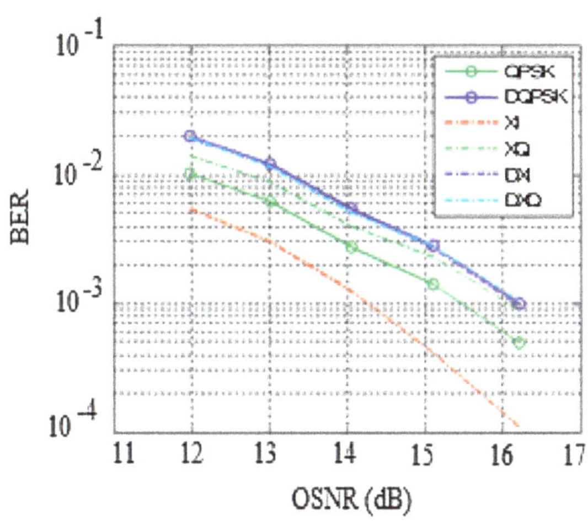

Figure 4.17. 2 Tbps channel performance measured by the variation of the BER versus OSNR; the modulation scheme is QPSK, with a 20 channel comb generator and 100 Gb s^{-1} per channel.

To demonstrate that the comb line sub-carriers can be stable and employable in Tera-bits s^{-1} optical transmission systems, we have measured the BER versus the OSNR of a probe channel, as shown in figure 4.17. The measured BERs for individual in-phase and quadrature phase components are also included in this figure. In order to save the cost of the field trial, we employ one I–Q modulator to modulate all sub-carriers at the same time, then one selected channel is optically filtered out and a probe channel is injected. The source used in the probe channel is also derived from a sub-carrier of the comb generator. The total number of channels is 20 with 100 Gb s^{-1} per channel, which is modulated by the QPSK modulation format and two multiplexed polarizations with a baud rate of 28 GBd. The noise is injected into the receiving sub-system via an optical coupler, not shown in the diagram of figure 4.14 but shown clearly in figure 4.11 with the unit ASE. The triangular points are measured under no error coding and the circles are obtained when error coding is employed. We observe no degradation of the performance for all sub-carriers of the generated comb. The linewidths of these comb sub-carriers were also measured and they match that of the original external cavity laser (ECL). The phases of these comb sub-carrier lines are indirectly interpreted in the electronic digital domain when processing the modulated optical signals of each modulated channel shown in figure 4.10(b). The OSNR of 16 dB is about 3 dB higher than the level required for a single channel. This is expected as there may be some crosstalk from the left and right adjacent channels of the probe channel.

4.8 Remarks

This section has presented the generation of a comb source whose spectra can be controlled by modulating the amplitude of the lightwave circulating in a re-circulating fiber loop. Synchronization of the optical fields of the comb lines is achieved via the use of ISOM such that constructive interference of the lightwaves is achieved. Shift left, shift right or dual sideband shift have been proposed and experimentally demonstrated. Only SSB shift comb sub-carriers have been employed in the optical transmitters to generate the optical superchannels for Tbps transmission systems. The superchannels have been experimentally demonstrated over installed fiber transmission lines operating at aggregate bit rates of 1.0 Tbps and 2.0 Tbps. By modulating these comb lines and transmitting over long non-dispersion compensating optical fiber transmission lines, we have achieved acceptable error rates. A BER of 10^{-3} is considered to be error free when error coding is used. Thus the linewidth of the comb generator can be considered to remain the same as that of the original external cavity laser. Currently, we are conducting extensive characterization of the sub-carriers in terms of linewidth and tenability of their spectral positions. Our current works are expect to employ these comb generators with enhanced bit rates at 400 Gb s^{-1} so that 10 or 20 × 400 G can also be demonstrated by using the high level modulation format, e.g. 32QAM or M-ary-QAM instead of QPSK, so that 10 Tbps can be generated and transmitted and received with acceptable BER. Furthermore, we have reported a dual band frequency shifting comb generator employing a dual re-circulating frequency shift ring in which two

ISOMs are incorporated and inline optical amplifiers operating in saturation mode. However, no modulation and transmission of these sub-carriers has yet been done. It is planned to demonstrate optical transmission systems employing these ultra-wide spectrum sub-carrier superchannels in our future works. An optical phase locking may be needed for this application so that all local oscillation sub-carriers can be simultaneously locked onto a referenced phase as that of the transmitted super-channel. Finally, the reported comb generator given in this chapter can be potentially used as a multi-carrier local oscillator for mixing with modulated optical sub-carriers in multi-channel coherent reception sub-systems for Tbps optical transmission systems.

4.9 Multi Tera-bits s^{-1} optical access transport technology

Tremendous efforts have been invested to developed multi-Tbps over ultra-long and metro distances and access optical networks. With the exponentially increasing demand on data transmission, storage and serving, in particular for the 5G wireless access scenarios, optical Internet networking has evolved to data center based optical networks, requiring novel and economical access transmission systems. This section describes the following. (i) Experimental platforms and transmission techniques employing band-limited optical components operating at 10 G for 100 G based at 28 GBd. Advanced modulation formats such as PAM4, DMT, duo-binary, etc, are reported and their advantages and disadvantages are analyzed so as to achieve multi-Tbps optical transmission systems for access inter- and intra-data-centered-based networks. (ii) Integrated multi-Tbps combining comb laser sources and microring modulators meeting the required performance for access systems are reported. Ten sub-carrier quantum dot lasers are employed in association with wideband optical intensity modulators to demonstrate the feasibility of such sources and integrated microring modulators acting as a combined function of demultiplexing/multiplexing and modulation, hence providing compactness and economy of scale. Under the use of multi-level modulation and direct detection at 56 GBd an aggregate of higher than 2 Tbps and even 3 Tbps can be achieved by interleaved two comb lasers of 16 sub-carrier lines. (iii) Finally, the fundamental designs of ultra-compact flexible filters and switching integrated components based on Si photonics for multi-Tbps active interconnection are presented. Experimental results on multi-channel transmissions and performances of optical switching matrices and the effects on that of data channels are proposed.

4.9.1 Introductory remarks

The emergence of DC centric networks has exerted tremendous pressure on traditional optical networks to be flattened and simplified, as well as increasing the network capacity and the contents to be delivered to end users. The capacity of the access networks should reach several Tera-bits s^{-1} when 4K video for the high definition and ultra-fast (faster than fast) delivery over fixed lines or wireless mobile 5G networks that is expected toward the end of this decade. The evolutionary network is schematically shown in figure 2.1, in which several Peta-bits s^{-1} capacity

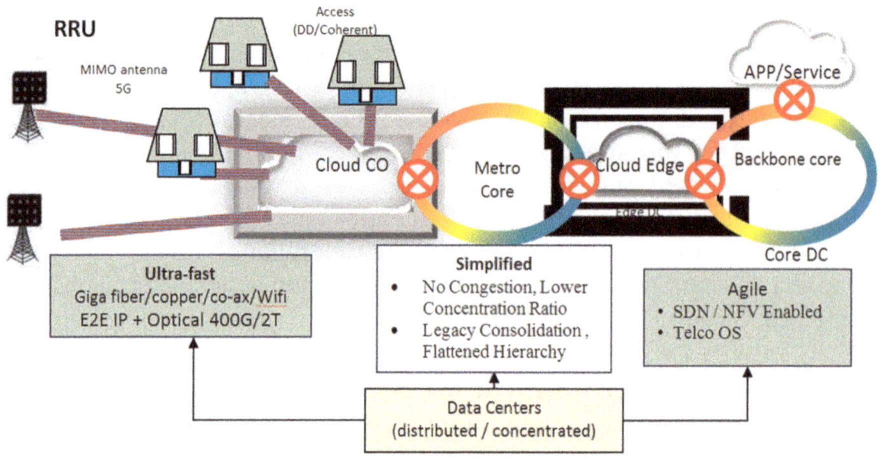

Figure 4.18. Evolutionary network modernization strategy. Cloud CO = central office cloud; DC = data center. 100 G access transmission systems.

would be transported in the long distance core and metropolitan core networks via several clouds at the edge as well as at local distributed nodes. The exchanges of traditional telecoms networks are expected to locate access clouds in their transformed flattened networks.

It is thus expected that multi-Tbps would be required to be deployed in the near future in the newly evolved networks. The economy of scale as well as the flexibility of such access transport networks or systems are very important factors so that bandwidth on demand can be offered to end users. This section thus addresses the following. First, the transmission and direct detection of four optical lanes of different and packed wavelength channels each carrying 25–28 Gbps depending on whether forward error coding (FEC) is required. Furthermore, several of such 100 G modules would result in accumulated capacity reaching Tbps. Second, we report on a baud rate of 28 GBd being employed to produce 56 Gbps via the modulation format PAM4, producing 56 Gbps per channel. Then we discuss the employment of comb sources and special modulators, such as microring types, to achieve multi-Tbps in both coherent and non-coherent reception. Finally we propose some integrated optical switching system to simplify the routing of ultra-fast optical channels to the end user locations (figure 4.18).

4.9.2 100 G access systems

A typical access transmission system is shown in figure 4.19 [32] in which a low cost VCSEL with a limited bandwidth of 18 GHz is used under a baud rate of 56 GBda and the data sequence is fed in to directly modulate the VCSEL. Shown also in this configuration is the T-bias and the PAM4 data sequence obtained SHF-10 001 attenuator 6 dB and a high speed combiner. The PAM4 eye diagram in the electrical domain is shown as the far left-hand side oscilloscope image in figure 4.19 followed

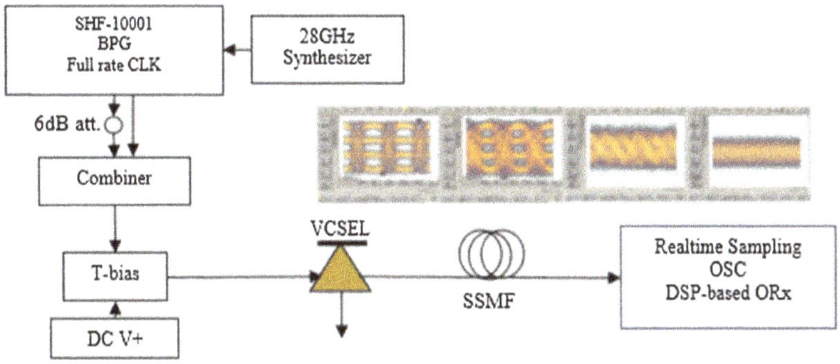

Figure 4.19. Typical access transmission system employing PAM4 direct detection and associated digital signal processing. SSMF = standard single-mode fiber; VCSEL = vertical cavity stimulated emission laser; DSP = digital signal processor; att = attenuator; CLK = clock.

by the optical output eye in back-to-back (B2B), transmitted through a length of standard single-mode fiber (SSMF), before feeding into the processor so as to evaluate the BER after passing through the minimum least square estimation (MLSE) algorithm. The performance of such a system is shown in figure 4.20. Under FECs of 20% and 7% we can expect that such a 56 G PAM4 can be transmitted over a few to some tens of kilometers of SSMF.

Alternatively, another low cost solution for access networks can be achieved using components developed for 10 G and operating at 28 G with DSP processing and pre-emphasis at the transmitter so as to produce systems of 100 Gbps via four optical wavelength transmitters employing an external modulation laser (EML) in which four lasers integrated with electro-absorption modulators. The transmission performance for such optical transmission is shown in figures 4.20(b), (c). It is believed that the footprints of such transponders employing 56 Gbps (PAM4) or 4 × 28 Gbps (NRZ) are the lowest cost to deploy into the access networks for DC centric networks or flattened traditional telecoms networks.

4.9.3 Tbps access transmission and routing technology

It is well known that the limitation of the electronic components is around 56 Gbps with bandwidth to around 35 GHz, in particular when digital signals are generated such as in high speed DAC and ADC. The noise contribution in these signal generators must also be considered as they grow as a cubic function of the frequency. This noise can then reduce the allowable OSNR, thus the amplitude levels when multi-level modulation format is employed. Multi-level modulation is required in order to increase the bit rate or the number of bits per symbol so that 100 G or 400 G can be obtained. In this section the 112 Gbps via 56 GBd PAM4 modulation format signals are described. Furthermore, in order to achieve the Tbps superchannels required, parallel channels need to be are simultaneously generated and transmitting. We propose here the use of comb laser sources and equivalent modulators and multiplexers via the device known as the microring modulator (MRM).

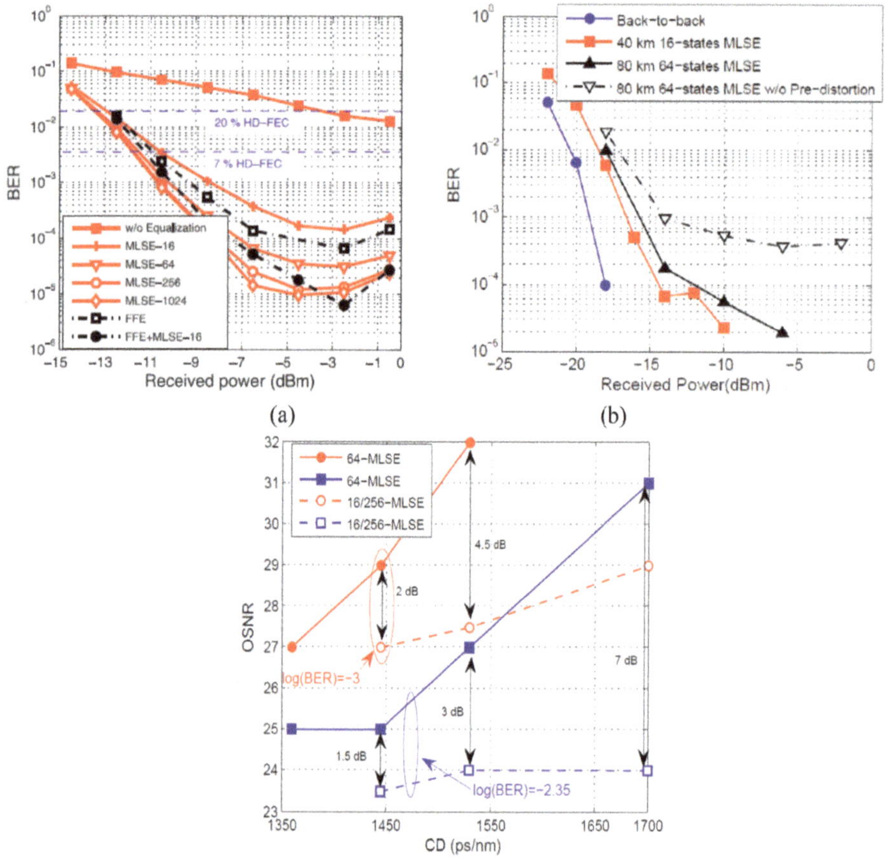

Figure 4.20. Performance of a low cost transmission system using band-limited optical components for metro access: (a) BER versus received optical power for PAM4 28 GBd; (b) BER versus received optical power for NRZ 28 G using 10 G components; and (c) OSNR versus residual chromatic dispersion in SSMF for NRZ 28 G.

4.9.4 Multi-carrier comb source direct detection systems

The MRM can act as an optical filter as well as a modulation device. In comb generation one can generate multiple sub-carriers from one original source in the frequency domain or in the time domain with a very short pulse repetitive sequence followed by a periodic Fabry–Perot filter whose free spectral range meets the required frequency spacing in a DWDM grid spectrum (figure 4.21). In the case of flex grid networks the periodic spaced grid can be occupied differently in multiple equally spaced wavelength grids. Figure 4.22 shows the paths for generation of comb lines in these two domains. Furthermore, there are a number of techniques for generating comb lines in the frequency domain via different optical processing structures including: (i) frequency shifting re-circulating optical loop with single sideband (SSB) shift of the guided lightwaves in the circulating cavity; (ii) phase modulation and carrier suppression return zero for shifting and then generating;

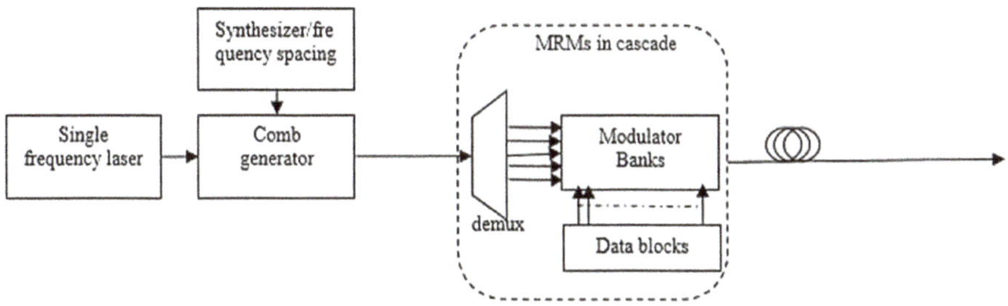

Figure 4.21. Schematic of multi-Tbps optical transmitter.

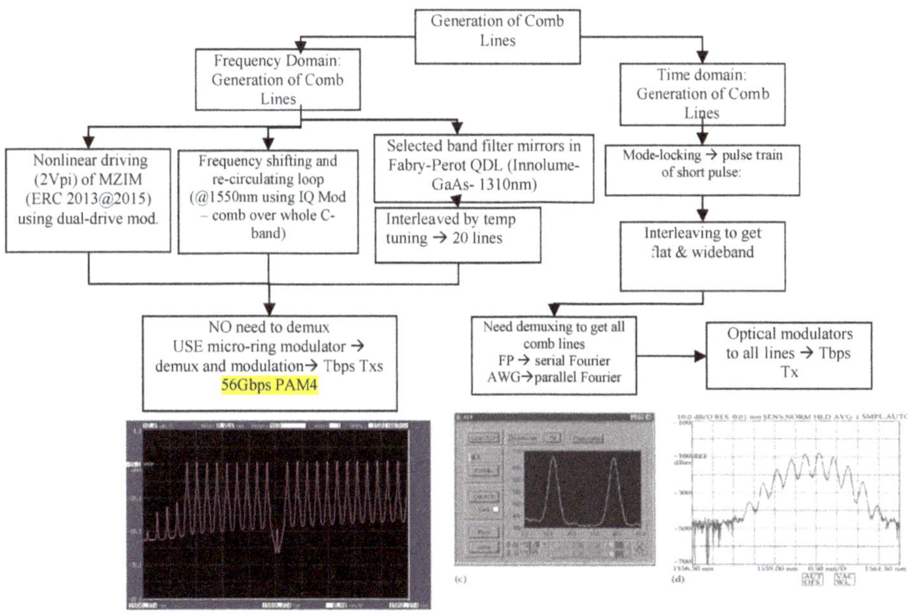

Figure 4.22. Methods of generation of comb sources in the frequency and time domain.

(iii) modulation integrated in a laser cavity; and (iv) short pulse generation and then super-continuum via a crystal waveguide and then filtering of determined frequency spacing by a Fabry–Perot and then optical amplification to the desired level of the comb lines.

Figure 4.23(a) to (c) display the optical spectra of the comb laser. Figure 4.23(a) shows the five comb lines under no modulation, while figure 4.23(c) depicts those comb lines under modulation by a digital data sequence. The five-sub-carrier comb source if modulated to produce 400 Gb s^{-1} per line then the aggregate bit rate would be 2.0 Tbps. We have demonstrated that by shifting left and shifting right the comb lines can cover the whole C-band with frequency shifting of 50 GHz which is one frequency shift applied by the frequency synthesizer to the optical modulator incorporated inside the re-circulating loop, as shown in figure 4.21.

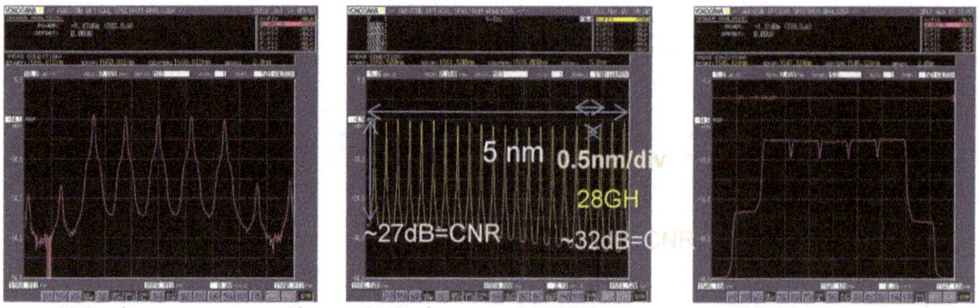

(a).Comb spectrum generated by (b).Comb spectrum by SSB frequency (c). Modulated spectrum of (a).

Figure 4.23. Comb spectrum generated by: (a) frequency shift re-circulating loop; (b) phase and amplitude modulation in nonlinear overdrive region; and (c) modulated spectrum of the comb laser of (a).

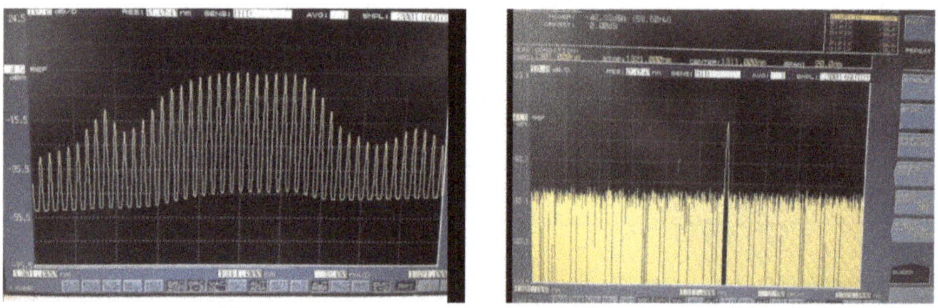

(a) (b)

Figure 4.24. Spectrum of a comb laser at the 1310 nm region: (a) all comb lines and (b) a filtered comb line.

In access networks the 1300 nm region can be employed in order to avoid the dispersion effects of the SSMF. One such comb laser in 1310 nm is shown in figure 4.24(a). The comb laser is fabricated using quantum dot structures in InGaAs with a flat broadband mirror at the two ends of the laser cavity[4]. A comb line can be filtered as shown in figure 4.24(b). Such a comb laser in the 550 nm region can be generated by an embedded modulator in a quantum dot laser cavity, as reported in [33, 34]. These comb lasers would offer the generation of modulated channels reaching a total capacity of 9.4 Tbps over the entire C-band with 100 Gb s^{-1} per channel and 94 sub-carriers of 50 GHz spacing. This band has now been extended to 94 wavelength carriers as optical amplifiers have been available and optimized over this C-band.

Employing the 1310 nm comb laser produced by Innolume GmbH and modulating using an external modulator with a 3 dB bandwidth of 35 GHz with the modulation format PAM4, we obtain 112 Gb s^{-1} per carrier, so the total aggregate capacity is 13×112 Gb s^{-1}, or 1.45 Tbps can be produced by this comb laser. The 1310 nm region ensures that the transmission is attenuation limited and not

[4] Innolume GmbH, Dortmund Germany; http://innolume.de/_pdfs/Comb/LD-1310-COMB-8.pdf.

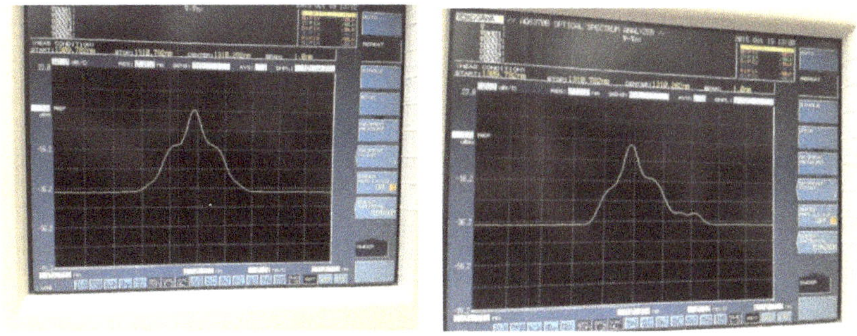

(a) (b)

Figure 4.25. Modulated spectrum of PAM4 56 GBd: (a) filtered at the center of the spectrum and (b) misaligned filtered modulated spectrum at PAM6 56 GBd.

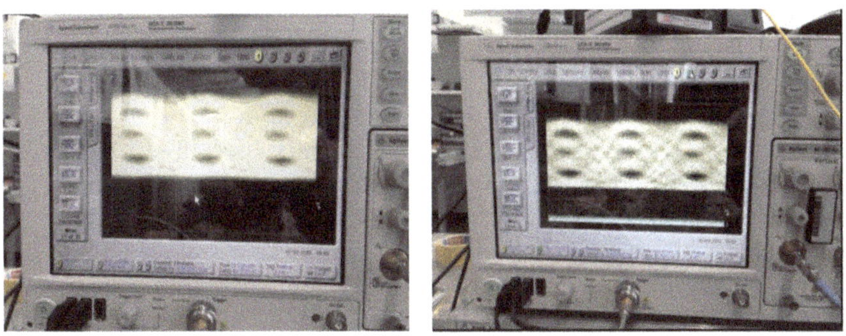

(a) (b)

Figure 4.26. Eye diagram at the output of transmitter PAM4 56 GBd of a filtered channel of the comb modulated channels.

dispersion limited. Figure 4.25 shows the spectra of the modulated comb line under matching of the filter center wavelength with that of the carrier comb line. The PAM4 eye diagrams under B2B are shown in figure 4.26. The BER is estimated at 1e-5. After 2 km transmission this is about 1e-3 under equalization using MLSE. The BER is similar for all 13 comb lines, indicating that the chromatic dispersion is not critical for this Tbps transmission system. Thus they are most suitable, with the employment of FEC, for low cost Tbps access or DC environments.

4.10 Tbps coherent reception systems

The coherent transmission technology using polarization multiplexing (PDM) and M-ary-QAM has been deployed throughout the long distance and metropolitan core networks. The flattening of telecoms networks under multi-Tbps has put tremendous pressure on network carriers to transform several fiber rings topologies into a mesh topology in the metro-core networks as close to the access users as possible. Although it is more complicated in terms of additional optical hybrid couplers

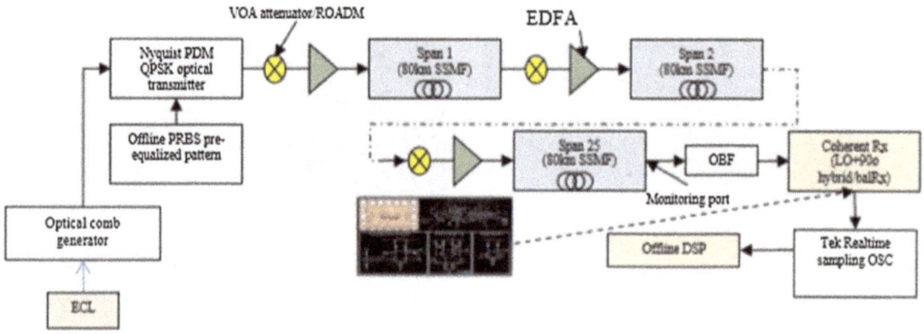

Figure 4.27. Schematic structure of the Nyquist PDM-QAM optically amplified multi-span transmission line. The inset is the advanced differential TIA optoelectronic receiver using CMOS 90 nm technology.

and local oscillators in the coherent reception sub-systems, it offers much better operating margins and a higher capacity, which is at least doubled by the PDM. It is also more expensive, but this cost is expected to be decreased with the integrated Si photonic technology under current intense development and research[5] [34–37]. Indeed a number of significant integrated photonics based on Si CMOS technology are now available commercially.

This section provides an overview of our demonstration of a multi-Tbps transmission system based on coherent transmission and reception techniques. A general multi-Tbps optical experimental platform is shown in figure 4.27. The comb laser source is generated using the frequency shift circulating technique described above. The comb source is demultiplexed into individual comb lines which are then fed into a pulse shape Nyquist transmitter through which the Nyquist sampling rate and roll-off are enforced on the optical modulator so that the modulated spectrum can reach close to a sharp roll-off. In this way the optical channels can be packed close together to minimize the crosstalk as well as be efficient in the spectrum, hence providing higher capacity for all channels transmitted over the C-band. An arrangement of the PDM-QAM transmitter is shown in figure 4.28. The spectrum of the 2 Tbps superchannel is shown in figure 4.29.

Alternatively, a microring modulator in a Si integrated photonic platform can be used as the Tbps transmitter, as shown in figure 4.30. The transmission performance of such a channel over 3500 km is shown in figure 4.29 and its transmission performance over 1600 km of optically amplified SSMF spans is shown in figure 4.31(c). A typical spectrum and eye diagram of the Nyquist channel are shown in figure 4.31(a), (b). The transmission of Tbps over this length allows shorter transmission with a number of reconfigured add/drop mux (ROADM), as expected for metro networks, in particular add/drop into access networks. It is noted that the coherent reception sub-system consists of an optical hybrid coupler through which the mixing of a comb laser acts as the multiple line local oscillator beating with all received channels in the photodetector after the optical demux. The balanced photodetectors are connected to the differential TIA so as to obtain the highest

[5] Yole Inc. 'A view on the Silicon Photonics market', report 2014.

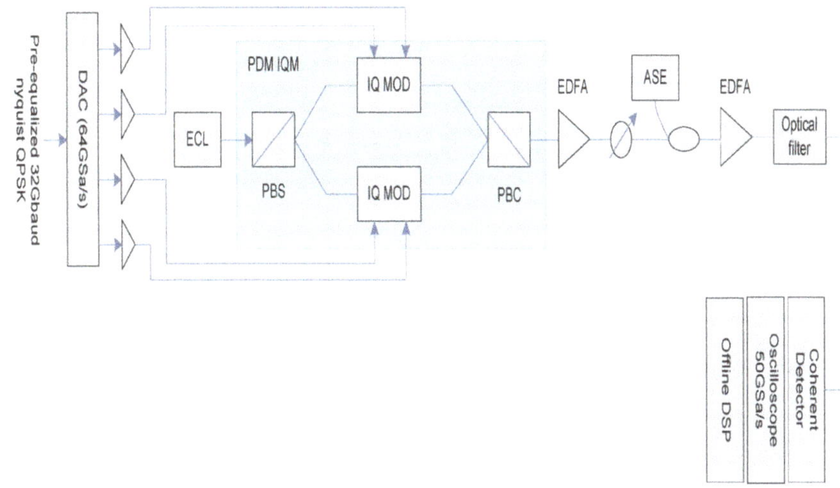

Figure 4.28. Experimental set-up of a 128 Gb s^{-1} Nyquist PDM-QPSK transmitter and B2B performance evaluation (see also figure 4.14).

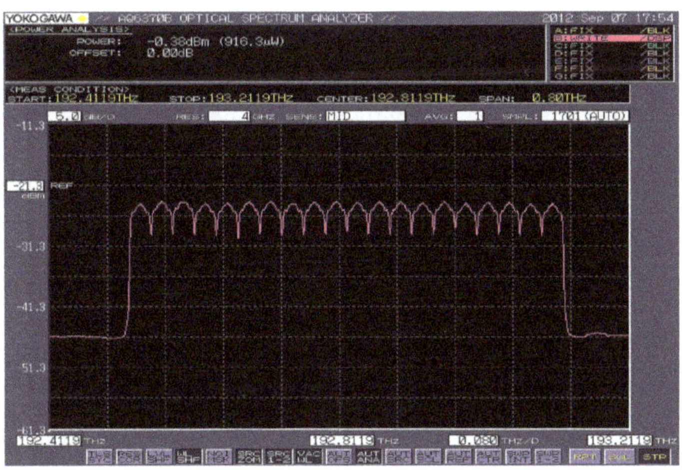

Figure 4.29. Spectrum of 2.0 Tbps optical channels using a nonlinear comb generator with two probe channels and 18 dummy channels.

output voltage level signals which are compatible for feeding into the ADC and then to the sampling oscilloscope and digitally processed. The BER is sufficient for further FEC to achieve error free results. Such Tbps superchannels can be integrated to transmit over multi-core fibers to provide a total aggregate capacity reaching a few peta-bps.

4.11 Optical interconnect for multiple Tbps access networks

Optically added drop mux for access networks must be available from the cloud to end users in multi-Gbps per user or even several Tbps to MIMO antennae for the

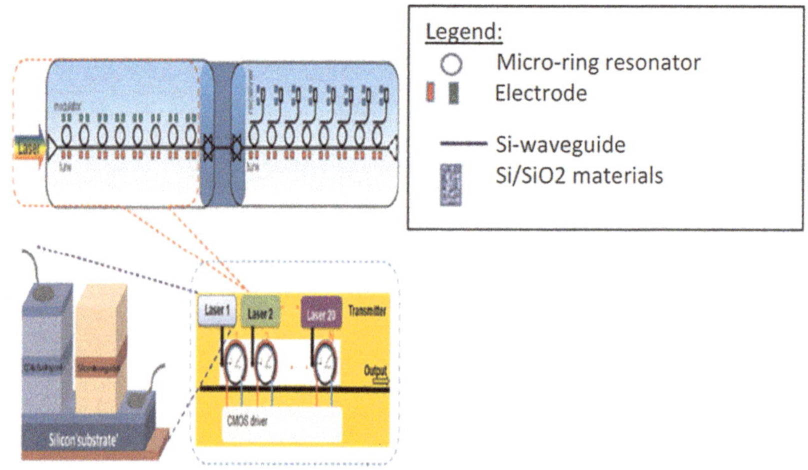

Figure 4.30. Tbps transmitter employing a microring modulator (MRM) as a simultaneous mux/demux and modulation device.

future 5G wireless access networks. We assume here that the multi-Tbps access networks accept channels of at least 100 Gbps to 400 Gbps and the superchannel 1 or 2 Tbps to 10 Tbps. Thus the bandwidth of channels can be variable, hence the flexi-grid WDM. This type of network will be installed in the up-coming modernization of optical networking. Indeed, these networks are under deployment by carriers such as Telefonica, Deutsche Telecom, etc. We propose an all optical interconnection for a multi-Tbps capacity network between metro-core and metro access environments, as shown in figure 4.32(a). It consists of an optical kernel which is composed of a matrix of cross waveguides and microring modulators/switches which can achieve demultiplexing or multiplexing and routing as well. Thus lightwave paths can be routed and wavelength channels can be selected. Such wavelength switching and routing in three directions is shown in figure 4.32(b). This structure can be implemented in Si integrated photonic circuits with a density switching matrix of order 100×100 without many problems on optical waveguide paths apart from the electrical connections and heating tuning section.

4.12 Remarks

This section has addressed the issues of access technologies for network capacity reaching several Tbps. Such networks will soon be active, in 2020, when 5G mobile networks and associate clouds, e.g. cloud radio access networks for 5G networking (CRAN) and 4K video delivery throughout the global Internet. The transmission technologies for 100 G per channel and Tbps have been presented and it has been proven that they are appropriate for such multi-Tbps transport networks. Furthermore, we have proposed an optical interconnect structure for add/drop from the metro-core to access networks. This optical interconnect can accept and operate with channels employing coherent or non-coherent reception.

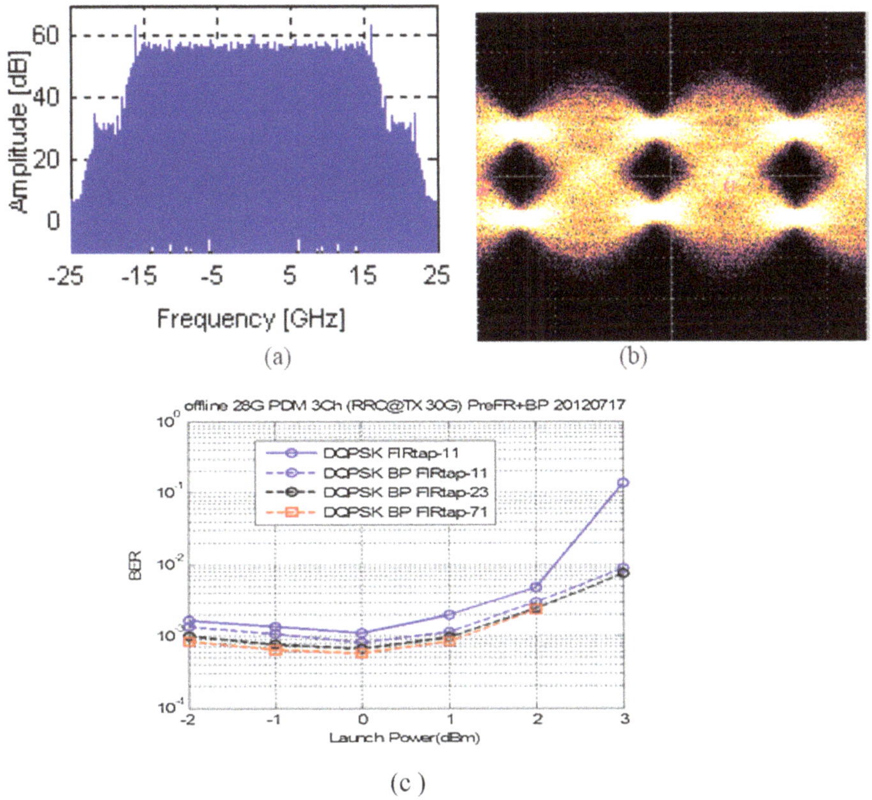

Figure 4.31. (a) Typical received spectrum, (b) Nyquist eye diagram and (c) BER versus launched power (dBm) of 20 channels with the central channel as the probe channel over a transmission distance of 1600 km non-DCF, 20 optically amplified 80 km SSMF spans with VOA inserted in front of EDFA. 28 GBd PDM-Nyquist QPSK: (a) single channel and (b) three channels.

Tbps transmission systems operating under coherent or direct detection are described using comb generators and multiple wavelength mux/demux and modulation. There is no doubt that such multi-Tbps access networks and DCN would be extensively deployed in the near future.

4.13 Low cost 1.6 Tbps using un-cooled comb sources

4.13.1 DSP-assisted Tbps low cost comb source system

For DCN and access networking with low cost and high capacity in the range of Tbps using coherent transmission techniques, the cost of the lasers are important to reduce the total cost. However, the processing power of the DSP can give us an innovative approach to mitigate any difficulties due to the fluctuation of the comb laser lines with respect to those of the carriers. This section gives a brief presentation of this technique. The set-up of this platform can be referred to figure 4.28, now depicted in figure 4.33, except that no Nyquist shaping is used so as to reduce the complexity of the DSP and hence the power consumption.

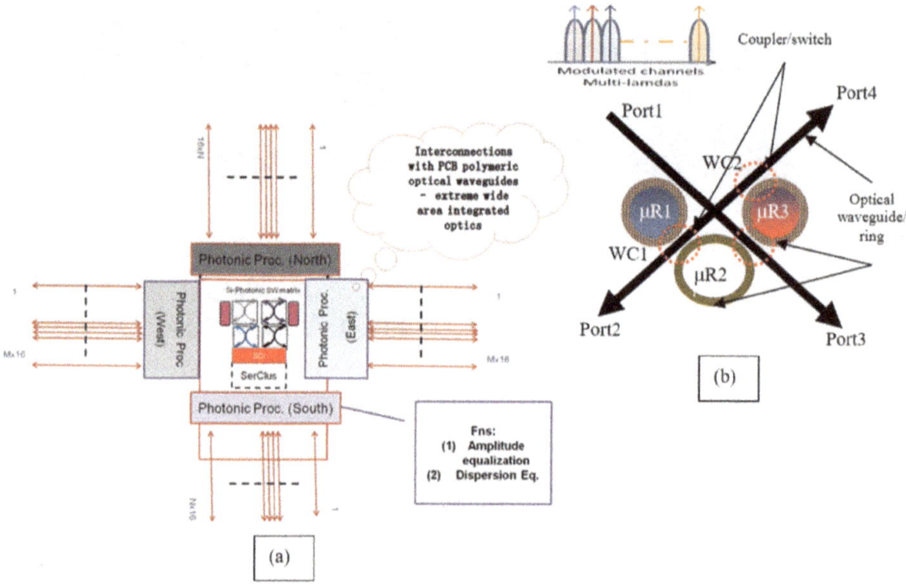

Figure 4.32. An optical interconnect consisting of an optical kernel and optical processor in a north–south–east–west arrangement, with fibers carrying 25 Tbps to 100 Tbps, all in the C-band with 92 wavelength channels or superchannels occupying multiple numbers of 50 G grid spacings.

The frequencies of the lasers of both the transmitter and local oscillator can be un-cooled and thus they can fluctuate slowly (figure 4.34). Consider the channels whose spectral distribution can be depicted in figure 4.35. Therefore, the processing of different channels is specific for individual channels. The proposed technique offers significant simplification of the receiver structure, hence lowering costs.

4.13.2 A simple generation of comb lines via cascade modulators

A novel technique to generate an ultra-flat and power efficient optical frequency comb (OFC) by serial cascading of electro-absorption modulators (EAM) and a single drive Mach–Zehnder modulator (MZM) has been shown in figure 4.36. In this proposed technique, the continuous wave (CW) light source is modulated and spectrum broadened by one electro-absorption modulator (EAM) and two MZMs, respectively. EAMs are in cascaded mode, producing ultra-short pulses followed by MZM, introduces intensity modulation and tuning the power variation of even and odd order sidebands at the same level for flattening the optical spectrum. Here, the first EAM acts as a sub-carrier generator, then the second EAM acts as a sub-carrier enhancer and the MZM acts as sub-carrier flattening. Using this technique, 63 sub-carriers with 10 GHz spacing and within 2 dB power fluctuation can be generated. The generated OFC has a bandwidth of 630 GHz. For theoretical and principal analysis of cascaded EAMs and MZM see [38]. This technique can generate multi-carriers with a uniform power level and

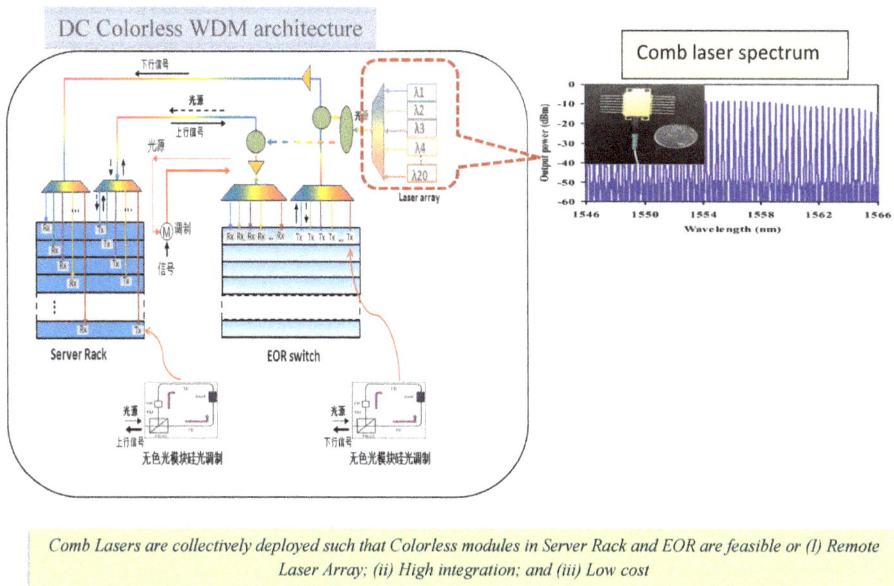

Figure 4.33. Multi-carrier single source comb laser and its use in both a transmitter and coherent reception local oscillator. The inset is the chip size of the laser for colorless WDM architecture for a DC.

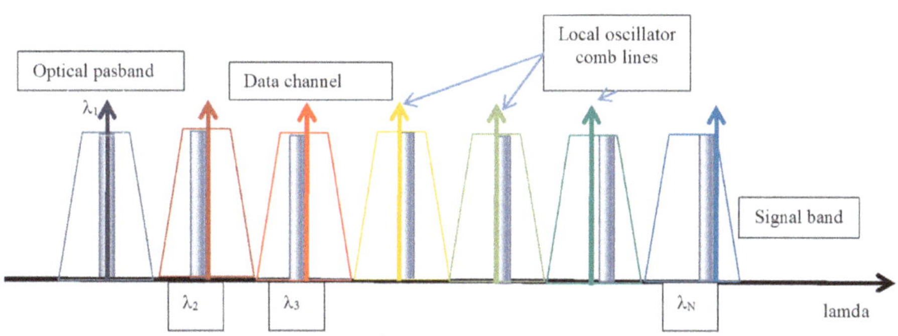

Figure 4.34. Spectral distribution of modulated channels, the comb lines of a multi-carrier laser and the optical passband of optical components.

reasonable high power spectrum which can be advantageous for dense wavelength division multiplexing (DWDM) and short distance multi-Tbps transmission for data communications [38].

4.13.3 Optical injection and comb generation

Generation of coherent multi-carrier signals by gain switching of discrete mode lasers has recently been reported [39] in which a DFB discrete mode laser is modulated to generate comb lines and in addition optical injection is also used to improve the linewidth of the comb lines. The linewidth measurement with

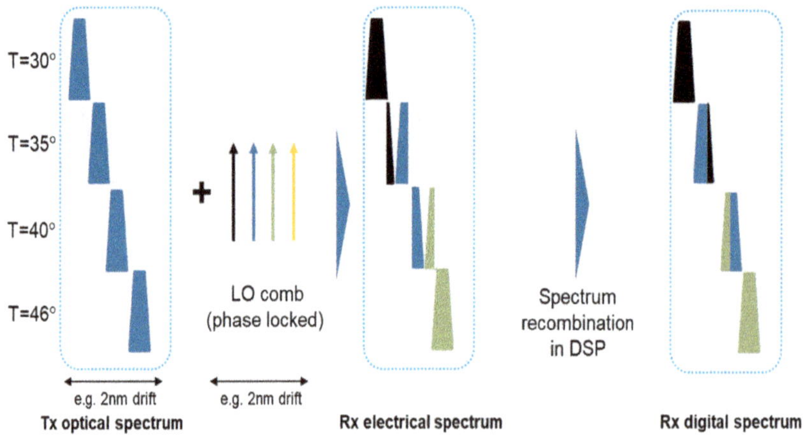

Figure 4.35. Spectra of channels in sequence of reception, mixing and DSP processing: after transmission from the optical receiver at different temperatures; local oscillator comb laser; electrical domain after mixing with LO drifted spectra; and after DSP processing or in the digital domain.

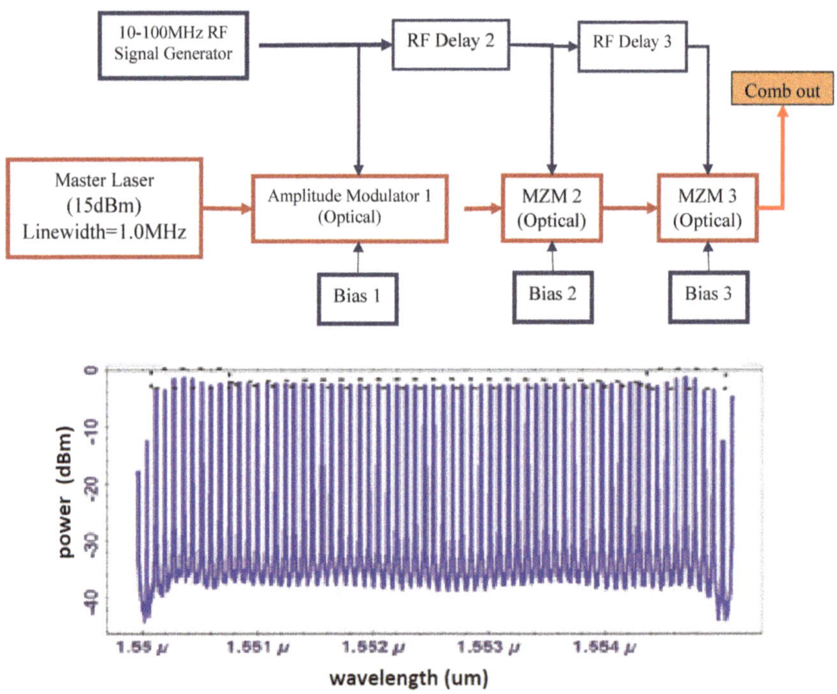

Figure 4.36. A simple comb generator by cascading three optical modulators driven into nonlinear regions by an RF signal generator whose frequency determines comb spacing.

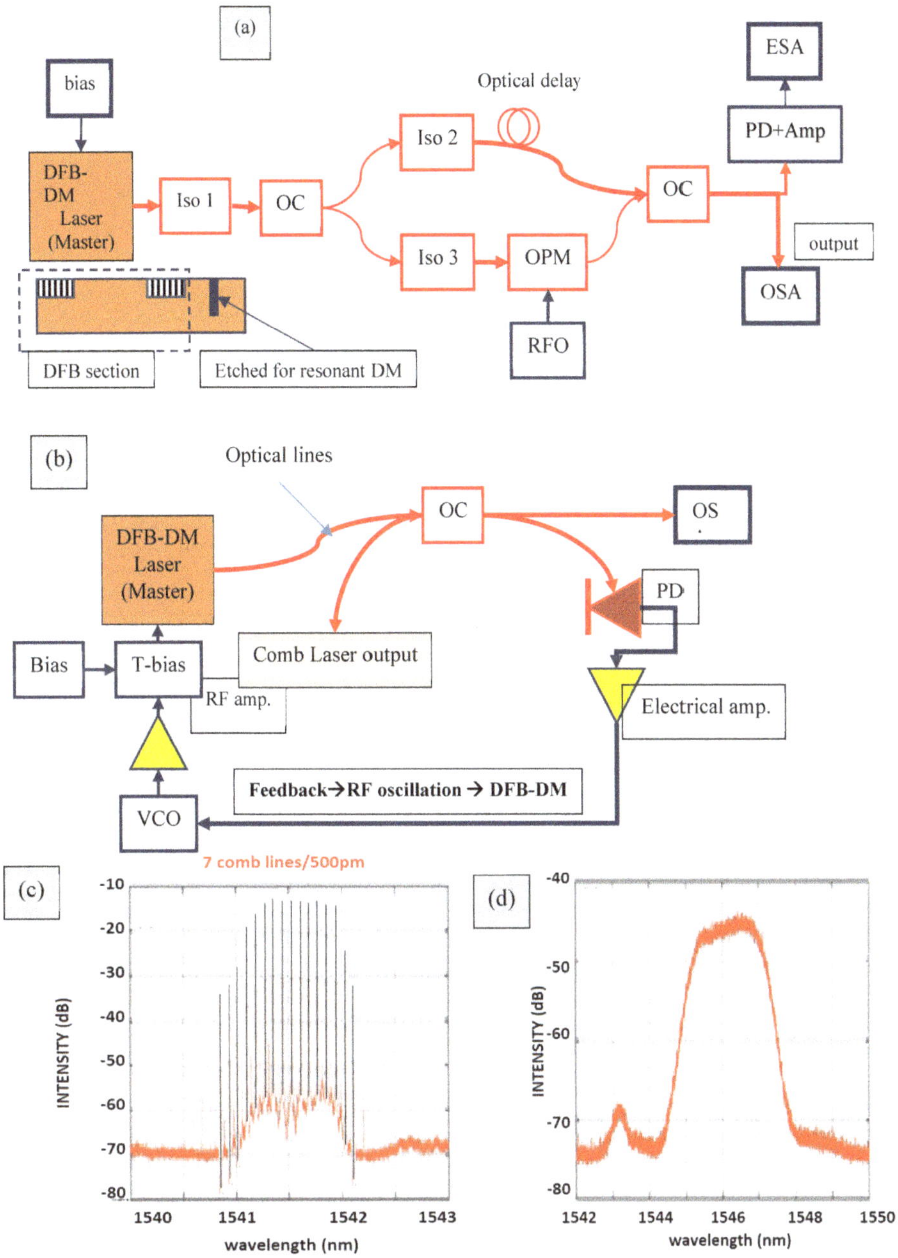

Figure 4.37. (a) Experimental set-up for linewidth characterization (plus insertion of the DFB-DM structure). (b) Experimental set-up for gain-switched DM and DFB lasers. (c) Optical gain-switched spectra for a DM laser generating comb lines. (d) Optical gain-switched spectra for a DFB laser. PM = phase modulator; RF = radio frequency; VCO = voltage control oscillator; PC = polarization controller; ESA = electrical spectrum analyzer; OSA = optical spectrum analyzer; Iso = isolator; OC = 3 dB optical coupler; OPM = optical phase modulator; RFO = radio frequency oscillator; ESA = electrical spectrum analyzer; OSA = optical spectrum analyzer; DFB-DM = distributed feedback laser incorporating the discrete mode by etching a Fabry–Perot laser cavity in the feedback path; PD = photodetector.

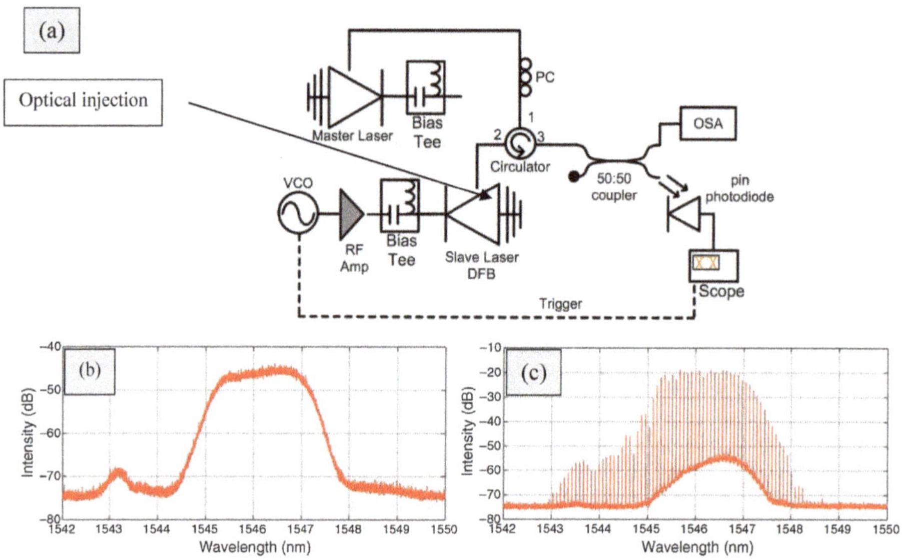

Figure 4.38. (a) Structure of an externally injected gain-switched DFB laser with external optical injection arranged in a master–slave coupled laser pair. (b) Gain spectra of the DM-DFB. (c) Comb line generation spectra of about 18 lines over a 2.0 nm window of the C-band 1546 nm.

self-homodyne coherent beating and the gain switching of discrete mode DFB can be used for generating comb lines. These set-ups are shown in figure 4.36.

The structure of the discrete mode (DM)-DFB laser shown in figure 4.37(b), which can be generating the multi-comb lines shown in (c) and gain spectra in (d). This DM-DFB is then improved by an external injection of a narrow linewidth laser.

The gain switching is achieved by applying amplified RF waves of a 10.7 GHz sinusoidal feature whose power output of 24 dBm in combination with a dc bias current about four times the threshold current about 200 mA to the laser via a bias tee. The optical output of the laser source is split using a 3 dB fiber coupler to enable simultaneous temporal and spectral measurements. The characterization of the multi-carrier signal is carried out using a high resolution (20 MHz) OSA and a high speed oscilloscope in conjunction with a 3 dB 50 GHz pin detector without any trans-impedance amplification (figure 4.38).

Compared to the comb laser whose comb lines are generated using dual optical modulators, as given in section 4.13.2, the comb line generation given in this section is much more compact and in complete integrated mode, thus less expensive and lower cost and applicable in low cost transmission links such as those employed in data center interconnections and access networks.

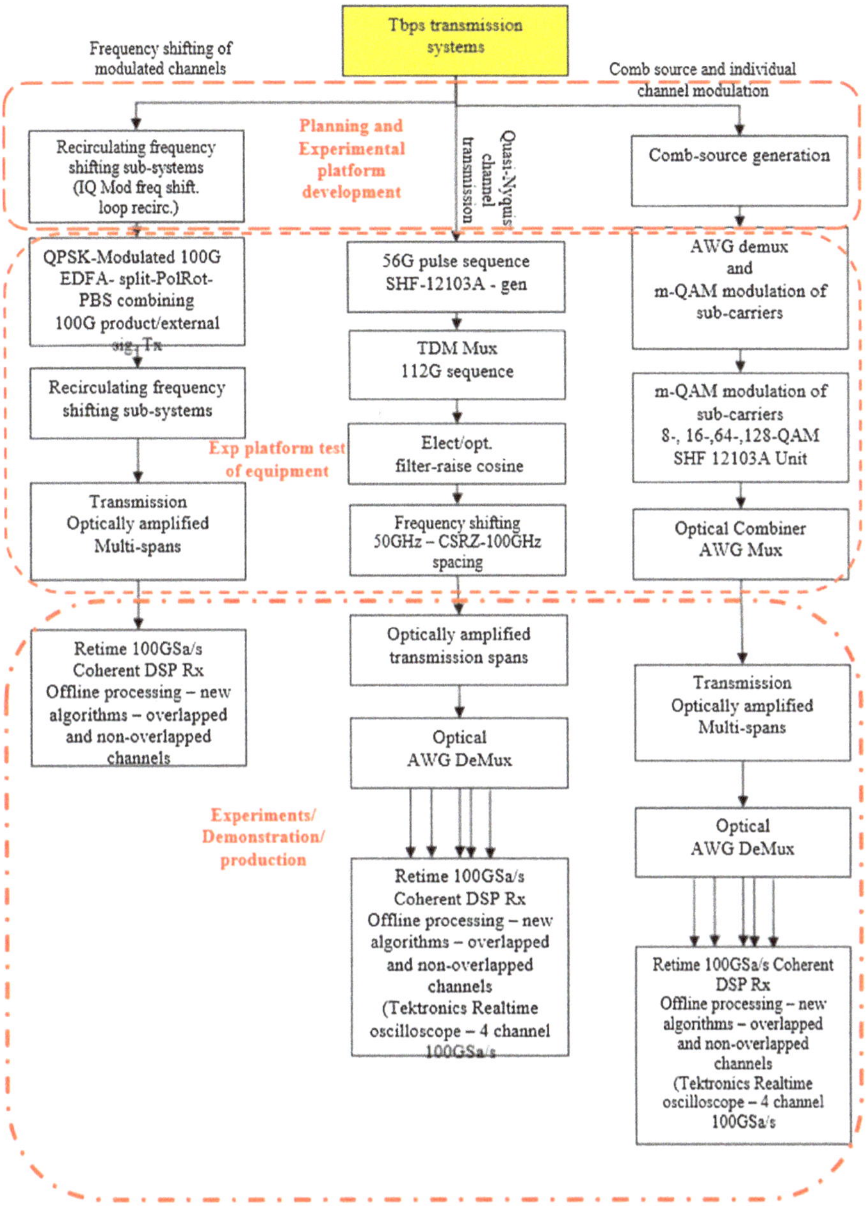

Figure 4.39. Schematic of the planning, experimental demonstration and production of a transport $N \times 100$ Gbps/carrier leading to multi-Tbps transport systems and their production.

4.13.4 Concluding remarks

This chapter has given a brief account of multi-carrier generation and their use in ultra-high capacity transmission lines for long distance, metropolitan networks and access networking, as well as in DC interconnection and inter-DC connections. The diagram depicted in figure 4.39 summarizes the engineering processes from the initial idea to planning and then production with credibility.

References

[1] Buchali F 2010 Technologies towards terabit transmission systems *Proc. European Conf. Opt. Comm. ECOC 2010 (Turin, Italy, September 2010)* paper We.6.C.1

[2] Stojanović N, Xie C, Zhao Y, Mao B, Gonzalez N G, Qi J and Binh L N 2013 Modified Gardner phase detector for Nyquist coherent optical transmission systems *OFC/NFOEC Technical Digest, OFC 2013 (Los Angeles, CA)* paper JTh2A.50

[3] Binh L N, Bangning M, Stojanović N, Xie C and Yang N 2013 Synchronous modulator incorporated re-circulating comb laser sources for Tbps superchannel transmission *Proc. OSA Congress, Conf. on Advanced Solid State Lasers (Paris, October 2013)* paper JTh2A.13

[4] Lundberg L, Karlsson M, Lorences-Riesgo A and Mazur M 2018 Frequency comb-based WDM transmission systems enabling joint signal processing *Appl. Sci.* **8** 718

[5] Hu H *et al* 2018 Single-source chip-based frequency comb enabling extreme parallel data transmission *Nat. Photonics* **12** 469–73

[6] Smith G H, Novak D and Ahmed Z 1997 Technique for optical SSB generation to overcome dispersion penalties in fiber-radio systems *Electron. Lett.* **33** 74–5

[7] Wei R, Yan J, Peng Y, Yao X, Bai M and Zheng Z 2013 Optical frequency comb generation based on electro-optical modulation with high-order harmonic of a sine RF signal *Opt. Commun.* **291** 269

[8] Mishra A K, Schmogrow R, Tomkos I, Hillerkuss D, Koos C, Freude W and Leuthold J 2013 Flexible RF-based comb generator *IEEE Photonics Technol. Lett.* **25** 701–4

[9] Xiao Y-q, Chen L, Li F and He H-z 2013 Novel full-duplex SSB WDM-RoF system with SLM technique for decreasing PAPR *Optoelectron. Lett.* **9** 309–12

[10] Xu Z, O'Sullivan M and Hui R 2010 OFDM system implementation using compatible SSB modulation with a dual-electrode MZM *Opt. Lett.* **35** 1221–3

[11] Zhang Y, O'Sullivan M and Hui R 2010 Theoretical and experimental investigation of compatible SSB modulation for single channel long-distance optical OFDM transmission *Opt. Express* **18** 16751–64

[12] Nazarathy M, Marom D M and Shieh W 2009 Optical comb and filter bank (de)mux enabling 1 Tb/s orthogonal sub-band multiplexed CO-OFDM free of ADC/DAC limits *Proc. ECOC 2009 (Vienna, Austria, 20–24 September)* Paper P3.12

[13] Zhang J, Yu J, Chi N, Dong Z, Shao Y, Tao L and Li X 2012 Theoretical and experimental study on improved frequency-locked multicarrier generation by using recirculating loop based on multifrequency shifting single-sideband modulation *IEEE Photonics J.* **4** 2249–63

[14] Li J, Li X, Zhang X, Tian F and Xi L 2010 Analysis of the stability and optimizing operation of the single-side-band modulator based on re-circulating frequency shifter used for the T-bit/s optical communication transmission *Opt. Express* **18** 17597–609

[15] Zhang J, Chi N, Yu J, Shao Y, Zhu J, Huang B and Tao B L 2011 Generation of coherent and frequency-lock multi-carrier using cascaded phase modulators and recirculating frequency shifter for Tb/s optical communication *Opt. Express* **19** 12891–902

[16] Zhang J, Yu J, Chi N, Shao Y, Tao L, Wang Y and Li B X 2012 Improved multi-carrier generation by using multi-frequency shifting recirculating loop *IEEE Photonics Technol. Lett.* **24** 1405–8

[17] Torres-Company V, Lancis J and Andrés P 2008 Lossless equalization of frequency combs *Opt. Lett.* **33** 1822–4

[18] Li J, Li X, Zhang X, Tian F and Xi L 2010 Analysis of the stability and optimizing operation of the single-side-band modulator based on re-circulating frequency shifter used for the T-bit/s optical communication transmission *Opt. Express* **18** 17597

[19] Binh L N 2011 *Guided wave photonics: fundamentals and applications with MATLAB®* (Boca Raton, FL: CRC Press, Taylor and Francis)

[20] Wooten E L *et al* 2000 A review of lithium niobate modulators for fiber-optic communications systems *IEEE J. Sel. Top. Quantum Electron.* **6** 69–82

[21] Campenhout J V, Green W M J, Assefa S and Vlasov Y A 2009 Low-power, 2 × 2 silicon electro-optic switch with 110 nm bandwidth for broadband reconfigurable optical networks *Opt. Express* **17** 24020

[22] Alferness R C 1979 Polarization-independent optical directional coupler switch using weighted coupling *Appl. Phys. Lett.* **35** 748–51

[23] Dedic I 2010 56 Gs s^{-1} ADC enabling 100 GbE *Proc. OFC 2010 Digital Transmission Systems, Optical Fiber Conf. (San Diego, CA, 21–25 March 2010)*

[24] Proakis J G 2000 *Digital Communications* 4th edn (Boston, MA: McGraw-Hill), section 9.2 p 560

[25] Christodoulopoulos K, Tomkos I and Varvarigos E A 2001 Elastic bandwidth allocation in flexible OFDM-based optical network *IEEE J. Lightwave Technol.* **29** 1354–66

[26] Uematsu† Y, Masuda A, Miyamura T and Hiramastu A 2012 Flexible virtualized optical transport networking technology *NTT Tech. Rev.* **10** 1–7

[27] Proakis J and Salehi M 2012 *Digital Communications* 5th edn (New York: McGraw-Hill)

[28] Binh L N 2008 *Digital Optical Communications* (Boca Raton, FL: CRC Press, Taylor and Francis)

[29] Proakis J G 2000 *Digital Communications* 4th edn (Boston, MA: McGraw-Hill), section 9.2 pp 554–65

[30] Bower P and Dedic I 2011 High speed converters and DSP for 100G and beyond *Opt. Fiber Technol.* **17** 464–71

[31] Kikuchi K 2011 Performance analyses of polarization demultiplexing based on constant-modulus algorithm in digital coherent optical receivers *Opt. Express* **19** 9868–80

[32] Karinou F *et al* 2014 Directly PAM4 modulated 1530 nm VCSEL enabling 56 G/λ data center interconnects *IEEE Photonics Technol. Lett.* **27** 1872–4

[33] Huhse D, Schell M, Bimberg D and Tarasov I S 2014 Generation of electrically wavelength tunable ($\Delta\lambda$ = 40 nm) singlemode laser pulses from a 1.3 μm Fabry–Perot laser by self-seeding in a fibre-optic configuration *Electron. Lett.* **30** 157–8

[34] Schmeckebier H, Fiol G, Meuer C, Arsenijević D and Bimberg D 2010 Complete pulse characterization of quantum-dot mode-locked lasers suitable for optical communication up to 160 Gbit s^{-1} *Opt. Express* **18** 3415–25

[35] Smith B T, Feng D, Lei H, Zheng D, Fong J and Asghari M Fundamentals of silicon photonic devices http://mellanox.com/related-docs/whitepapers/KOTURA_Fundamentals_of_Silicon_Photonic_Devices.pdf

[36] Gardes F Y, Thomson D J, Emerson N G and Reed G T 2011 40 Gb/s silicon photonics modulator for TE and TM polarizations *Opt. Express* **19** 11804–14

[37] Vlasov Y A 2012 Silicon integrated nanophotonics: road from scientific explorations to practical applications *CLEO'2012* plenary paper

[38] Ujjwal and Thangaraj J 2018 Generation of an ultra-flat and power efficient optical frequency comb by cascading of electro-absorption and single drive Mach–Zehnder modulator *Opt. Eng.* **57** 126106

[39] Anandarajah P M, Maher R, Xu Y Q, Latkowski S, O'Carroll J, Murdoch S G, Phelan R, O'Gorman J and Barry L P 2011 Generation of coherent multicarrier signals by gain switching of discrete mode lasers *IEEE Photonics J.* **3** 112–22

IOP Publishing

Transmission and Processing for Data Center Networking

Le Nguyen Binh

Chapter 5

Photonic signal processors

This chapter gives an overview of processing signals in the optical and photonic domains. In addition the devices and processors for the fifth-generation wireless networks (5G networks) which require substantial high capacity optical transport techniques are described. From this perspective we give again a brief description of optical transmission in the access layer.

5.1 Optical transformed channels and transmission: spectral domain processing

In processing there are two commonly known domains in electrical or photonic systems: the temporal and spectral domains (or time and frequency domains). This chapter deals with photonic processing in the spectral domain and temporal domain, including processing using neural optical networking.

5.1.1 Optical Fourier transform (OFT) based structure

A superchannel transmission system can also be structured using optical fast Fourier transform (OFFT) as demonstrated in [1] and shown in figure 5.1, in which MZDI components (see figure 5.1(a)) act as the spectral filter and splitter (see figure 5.1(b)), the optical FFT. The outputs of these MZDI are then fed into coherent receivers and processed digitally, as in figure 5.1(c), with the electro-absorption modulator (EAM) performing the switching function so as to time de-multiplex the ultra-fat signal speed to a lower speed sequence so that the detection system can decode and convert to the digital domain for further processing.

The spectra of superchannels at different positions in the transmission system can be very close to the rectangular shape as expected from the OFDM in the optical domain, that is, very similar to those obtainable in the electrical domain as given in figure 5.2 [2].

doi:10.1088/978-0-7503-2292-8ch5

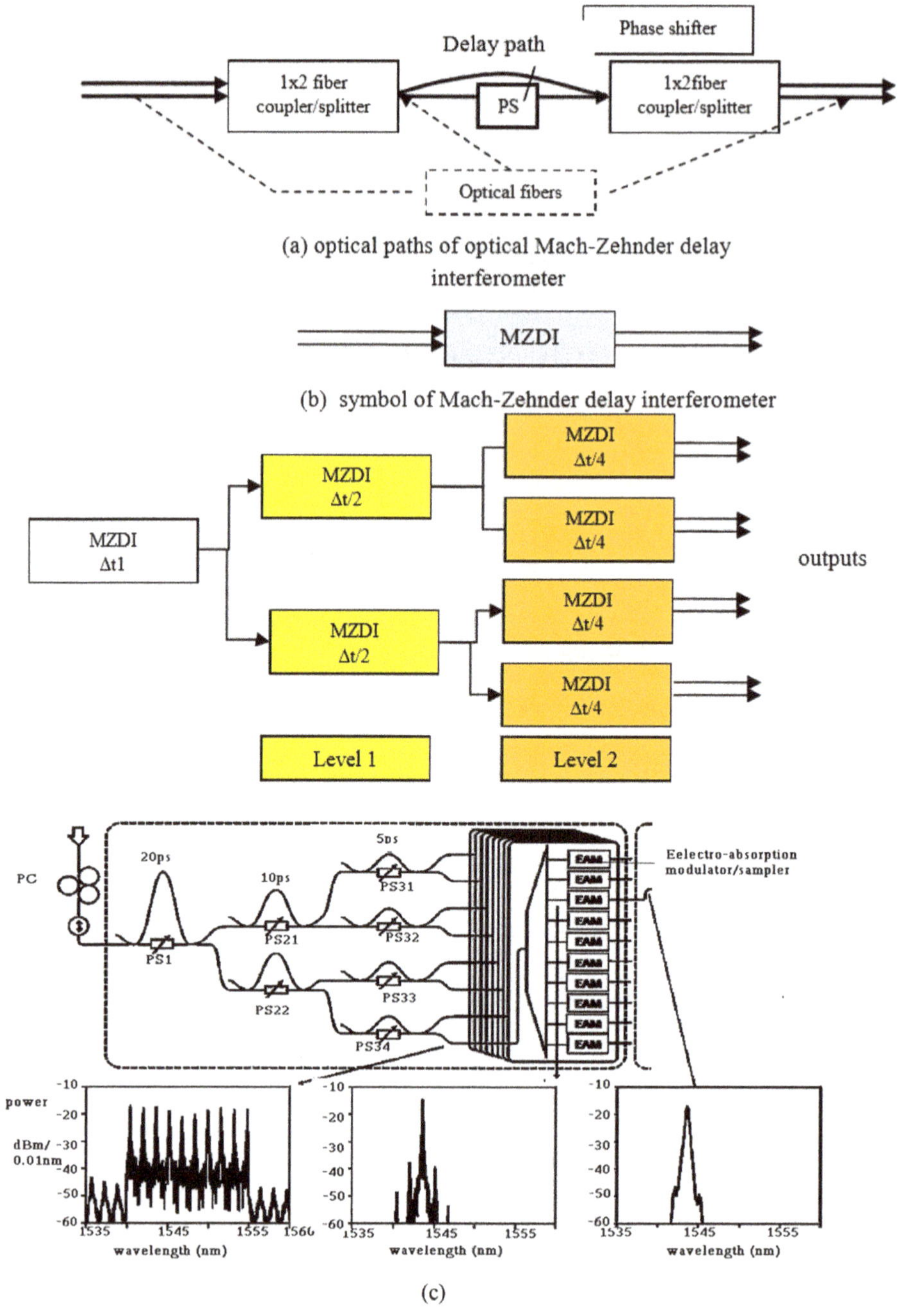

Figure 5.1. Operations by guided wave components using fiber optics. (a) Guided wave optical path of a Mach–Zehnder delay interferometer or asymmetric interferometer with a phase delay tunable by thermal or electro-optic effects. (b) Block diagram representation. (c) Implementation of optical FFT using the cascade stages of fiber optical MZDI structure (reproduced from [2]). EAM = electro-absorption modulator used for demultiplexing in the time domain. Note also the phase shifters employed in MZDIs between stages to tune the exact delay between two branches. The insets are spectra of optical signals at different stages, as indicated for the optical FFT (serial type). PC = polarization controller.

5.1.2 Optical Fourier processor OFT based structure

The processing of superchannels can be considered as similar to the digital processing of individual sub-channels except when there may be crosstalk between sub-channels due to overlapping of certain spectral regions between the considered channel and its adjacent channels.

Thus for the Nyquist QPSK sub-channel the DSP processing would be much the same as for QPSK dense wavelength division multiplexing (DWDM) for 112 G described above with care taken for the overlapping either at the transmitter or at the receiver. The error vector magnitude is a parameter that indicates the scattering of the vector formed by the I and Q components departing from the center of the constellation point. The variance of this EVM in the constellation plane is used to evaluate the noise of the detected states, hence the Q-factor can be evaluated with ease and thus the BER using the probability density function and the magnitude of the vector of a state on the constellation plane. The BER of the sub-channels of the OFDM superchannel is shown in figures 5.3(b) and (c) [1] for different percentage overloading due to FEC. The loading factor is important as this will increase the speed or symbol rate of the sub-channel one has to offer. The higher this percentage, the higher is the increase in the symbol rate, and thus high speed devices and components are required.

In the receiver of the optical OFDM superchannel system of [1], to judge the effectiveness of the optical FFT receiver, three alternative receiver concepts were tested for a quadrature phase shift keying (QPSK) signal. A QPSK signal is chosen because it was not possible to receive a 16QAM signal with the alternative receivers owing to their inferior performance [1]. First, a sub-carrier with a narrow bandpass filter is used to extract a sub-channel. The filter passband is adjusted for the best performance of the received signal (figure 5.3). The selected filter

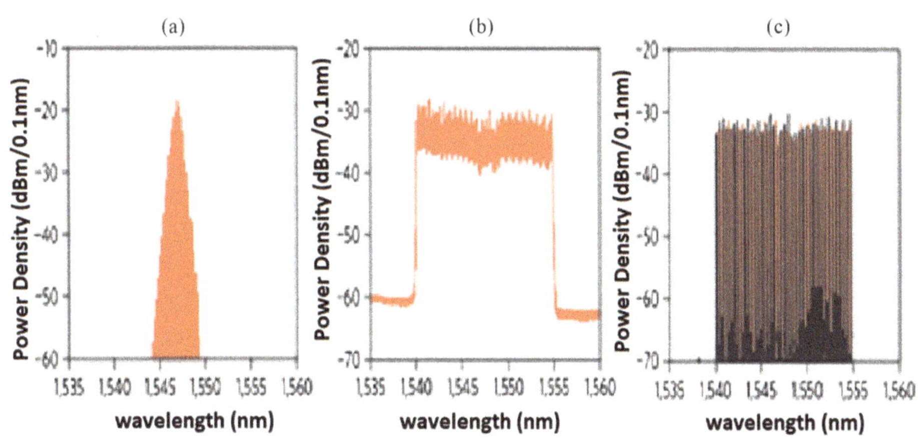

Figure 5.2. Spectrum of superchannels and demuxed channels. All sub-channels are orthogonal and thus the name OFDM (orthogonal frequency division multiplexing). (a) At output of transmitter, (b) after fiber propagation and (c) after polarization demux. Reprinted by permission from Macmillan Publishers Ltd: [1], copyright (2011).

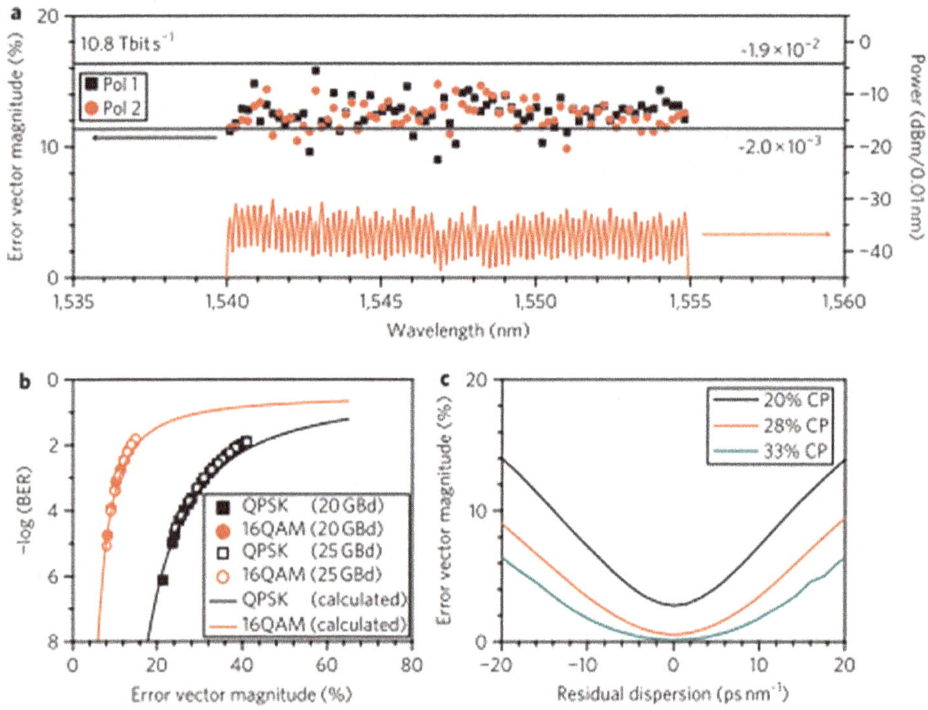

Figure 5.3. All-optical 10.8 Tbit s^{-1} OFDM results. (a) Measured error vector magnitude (EVM) for both polarizations (symbols) and for all sub-carriers of the OFDM signal decoded with the all-optical FFT. The estimated BER for all sub-carriers is below the third-generation FEC limit of 1.9e-2. The optical spectrum far left trace, is drawn beneath. (b) Relationship between BER and EVM. Measured points (symbols) and calculated BER as a function of EVM for QPSK and 16QAM. (c) Tolerance towards residual chromatic dispersion of the implemented system decoded with the eight-point FFT for a cyclic prefix of 20%, 28% and 33%. Reprinted by permission from Macmillan Publishers Ltd: [1], copyright (2011).

bandwidth is 25 GHz. The constellation diagram shows severe distortion. When using narrow optical filtering, one has to accept a compromise between crosstalk from neighboring channels (as modulated OFDM sub-carriers necessarily overlap) and inter-symbol interference (ISI) owing to the increasing length of the impulse response when narrow filters are used. Narrow filters can be used, however, if the ringing from ISI is mitigated by additional time gating. The reception of a sub-carrier using a coherent receiver is then performed. In the coherent receiver, the signal is down converted in a hybrid coupler, and detected using balanced detectors and sampled in a real-time oscilloscope. Using a combination of error low-pass filtering due to the limited electrical bandwidth of the oscilloscope and digital signal processing, the sub-carrier is extracted from the received signal. This receiver performs better than the filtering approach, but a larger electrical bandwidth and sampling rate of the ADC and additional DSP would be needed to eliminate the crosstalk from other sub-carriers and then to achieve a performance similar to that of the optical FFT.

Thus OFDM may offer significant advantages for superchannels but additional processing time would be required, while for Nyquist QPSK superchannels it allows better performance and less complexity in the receiver DSP sub-system structure.

5.2 5G optical transport networking: from photonic devices to processors

In this section we present the urgency of optical transport networking evolution for 5G delivery and ultra-broadband services to users and communities of users, in the following: (a) cloud core, metro-cloud core and edge cloud networking structure with optical-SDN and SDN/NFV; (b) photonic enabled technologies including principal devices and photonic processors; (c) security aspects and transmission technology for secret keys in co-transmission of massive data transport.

5.2.1 Introduction

This section gives a brief introduction on the infrastructure of 5G networks and its evolution over recent years. The evolution of such 5G networks is continuously happening depending on the technological development of equipment by manufacturers of 5G networking. We must outline this evolution herewith in this section to set the tone for such 5G networking.

5.2.2 5G infrastructure evolution

In recent years, an evolution of the traditional radio access network (RAN) with a base station (BS) at each cell tower site (1G/2G), where all the baseband and radio processing is integrated, to a separated baseband and radio part of the BS (3G/4G) into a baseband unit (BBU) and a remote radio head (RRH) (or RRU, unit) has occurred. Cloud-RAN (C-RAN), where the RRHs are connected to a BBU pool, is expected to play an important role in 5G networks, reducing CaPeX as well as improving performance and energy efficiency via coordinated multipoint (CoMP), capacity sharing and optimization. Mobile fronthaul is a key network element in the C-RAN architecture to connect centralized baseband units (BBUs) with RRHs. Mobile fronthaul will need to meet the strict requirements of the key technologies that will be the 5G enablers, such as massive MIMO, millimeter-wave and ultra-dense networks. Only the optical fiber based mobile fronthaul can meet the high capacity demands of 5G and thus 5G not only brings great opportunities but also great challenges for the optical communications industry. The potential revenue will be greatly affected by the architectures and interfaces that will be used and a careful investigation and roadmap strategy is necessary at this early stage of 5G deployment.

Data centers (DCs) and 5G will drive growth in the optical communications market, and the optical equipment market will further grow by ~4.5% toward 2022 accordingly (according to the latest forecasts). The driving forces for the growth of 5G will be the preparation for 5G/enhanced mobile broadband (eMBB) (phase I), the need for machine-to-machine (M2M) and Internet of Things (IoT) applications (phase II) in the optical fronthaul, the shift from data video to augmented reality/virtual reality (AR/VR), applications such as 3D-holographic imaging for consumers or vehicle-to-the-X (V2X), and smart factories/cities. Optical communications technology will be the enabling technology not only for fronthaul (FH) but also for backhaul (BH) and mid-haul (MH) (or X-haul), note also figure 5.4.

5.2.2.1 Evolutionary distributed DCs

It is expected that the optical transport networks will evolve to 5G cloud-based network topology and this can be visualized as shown in figure 5.5. The evolution can be expressed as a multi-step process as follows. (i) Massive evolution of the

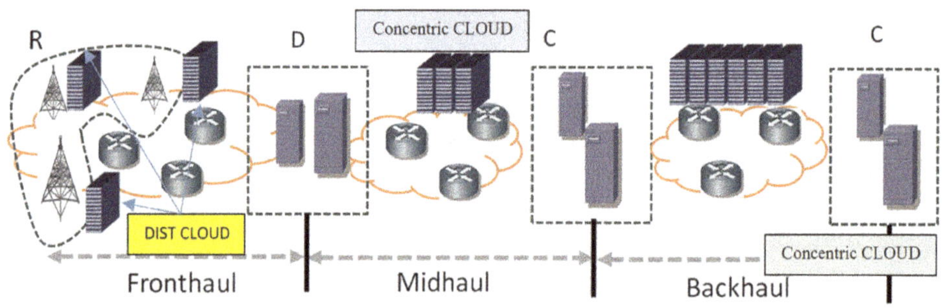

Figure 5.4. Optical communications and DC clouds in distribution and concentration architectures as the enabler for 5G X-haul.

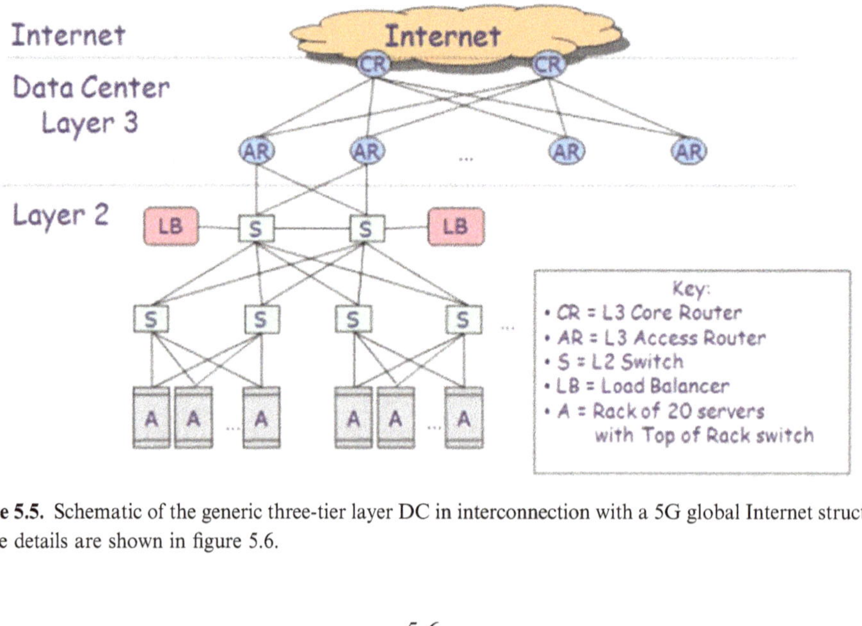

Figure 5.5. Schematic of the generic three-tier layer DC in interconnection with a 5G global Internet structure whose details are shown in figure 5.6.

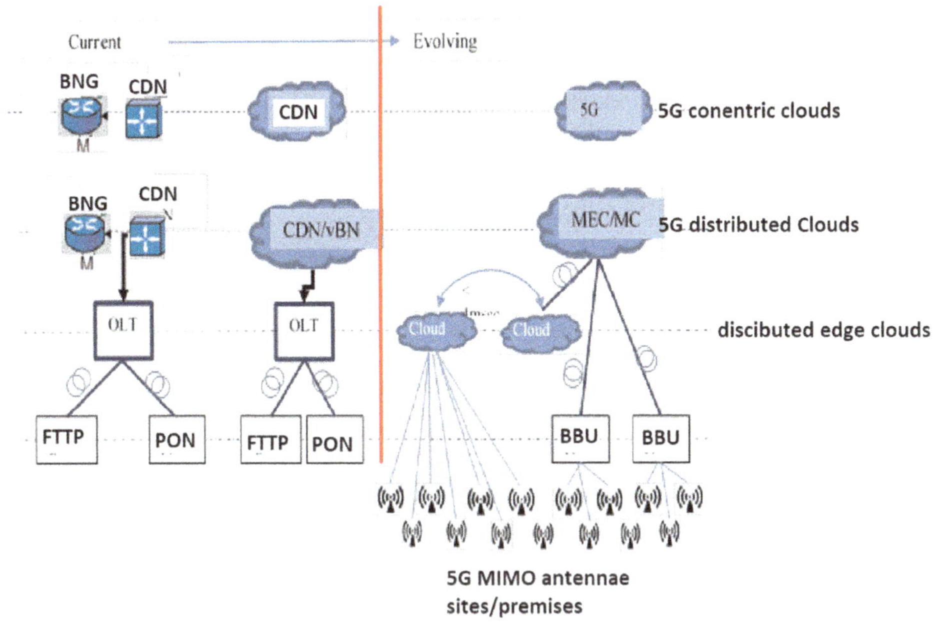

Figure 5.6. Evolution of optical transport networks to cloudified networks for 5G. CDN = content delivery networks; FBB = fixed broadband; CDN = current delivery network; vBNG = variable business network group; OLT = optical line terminal; MCE = mobile cloud engine; BBU = baseband unit; FTTP = fiber to the premise; PON = passive optical networks.

transport network from IP-based architecture of 4G to cloud-based architecture for 5G, driven by changes in ultra-high capacity and dynamic demand, the wireless and core network architecture of evolutionary traditional telecoms, and cloud DC networking. (ii) Aggregation of DC interconnect to metro-core networks (MCN). The low latency and wideband services of 5G will drive the down shifting of the backbone node to metro node. Alternatively, the whole terrestrial or intercontinental network can be viewed as a metro transport network in which low cost, low latency and ultra-high capacity must be met. These are the new challenges to be met by the DC networking infrastructure or telecoms carriers or effectively all 5G service providers (5 GSP). (iii) As far as the radio access networks (RANs), shown in figure 5.5, are to be densely distributed, metro distributed RAN (D-RAN) layers are formed and flattened on which the cloud data units and centralized units can be aggregated.

Cloud-RAN can enhance the gain of a base station in a 4G network by 10 dB based on test results in Japan and more obviously in a 5G environment. This thus requires a unification of the CPRI and eCPRI. The optical line terminals (OLTs) are to evolve into cloud-based mobile engines (MCEs) and receiving or transmitting to new 5G core nodes, the cloud-based topological network with several OTTA and mmW MIMO antenna sites. 5G network architecture is trending towards common and dynamic broadband and metro private line networking as well as multi-service transport.

5.2.3 Optical transport networking evolution for 5G delivery

5.2.3.1 State-of-the-art, current challenges and the role of novel technologies

Currently the radio-optical medium CPRI is the standard interface carrying the data between the RRH and the BBU pool. In the CPRI interface a wireless signal is digitized using a 15 bit resolution for each I/Q component of the complex waveform. For a two sector 4×4 MIMO channel of 20 MHz (4G), based on the CPRI-7-option and a sampling rate of 30.72 MS s^{-1}, a bit rate as high as 9.8304 Gb s^{-1} is required. For a potential 5G scenario, as an example 200 MHz bandwidth, 64×64 M-MIMO, and three sectors, there will be a need for a 240 (!) CPRI-7 like requirement (equivalent to 2.4 Tb s^{-1}). As a result bandwidths (BWs), well above 100 Gbps might be required from the optical network. Disaggregated RAN is an evolution of C-RAN to relax the bandwidth requirements at the fronthaul using functional splits. eCPRI is an enhanced version of CPRI to support more efficient interfaces as an onset for 5G development. eCPRI can support 5G 64×64 MIMO scenarios (25 Gb s^{-1}). The eCPRI interface specification can be supported by ethernet-switched or IP-routed fronthaul networks, or similar types of transport networks. However, different technical challenges emerge depending on where the split is, leading to a trade-off between fronthaul, BW versus latency requirements, complexity and cost of RRH. In particular, fronthaul latency needs to meet the requirements for the end-to-end latency on a 5G network which can be in the order of 10 ms (cloud assisted driving) down to 0.25 ms (high-frequency trading). A higher layer split (HLS) has already been agreed, but no consensus has been reached (at the time being) for a lower-layer split (LLS).

Different optical network solutions/interfaces can be applied to meet the 5G challenges at the fronthaul, such as high capacity and flexibility and also latency and low power consumption. In all cases, current network evolution and optimization is needed. An optical transport network solution (OTN/WDM) is a robust solution which can provide reliable and broadband fronthaul delivery to eMBB, M2M/IoT and the cloud using high capacity DWDM. OTN is simple, reliable and transparent for CPRI/eCPRI, however, an evolution to the access ring is needed, or relevant approaches (figure 5.7) need to be investigated. Ethernet (Eth-RoE and IP/Eth-eCPRI) can be a promising option to support functional splits. Packet switched networks based on ethernet seem to be a promising approach for 5G as they offer resource sharing, and are reconfigurable.

Ethernet is a standard technology, widely adopted, and can be easily upgraded to include software defined networking and network function virtualization (SDN/NFV) aspects. However, low delay requirements must be met in all cases e.g. packet delay and variation is dependent on traffic load, and synchronous ethernet will be needed. Last but definitely not least, a passive optical network (PON) can reduce the cost of fiber deployment and reuse the current optical distribution network (such as OTN/WDM) using SDN/NFV to optimize the network according to applications. However, PON capacity needs to be upgraded (25 G to 50 G or 40 G/NGPON2); latency issues need to be met using a low latency dynamic bandwidth allocation (DBA) algorithm and protection-multi-wavelength allocation with e.g. NGPON2 might be required. High

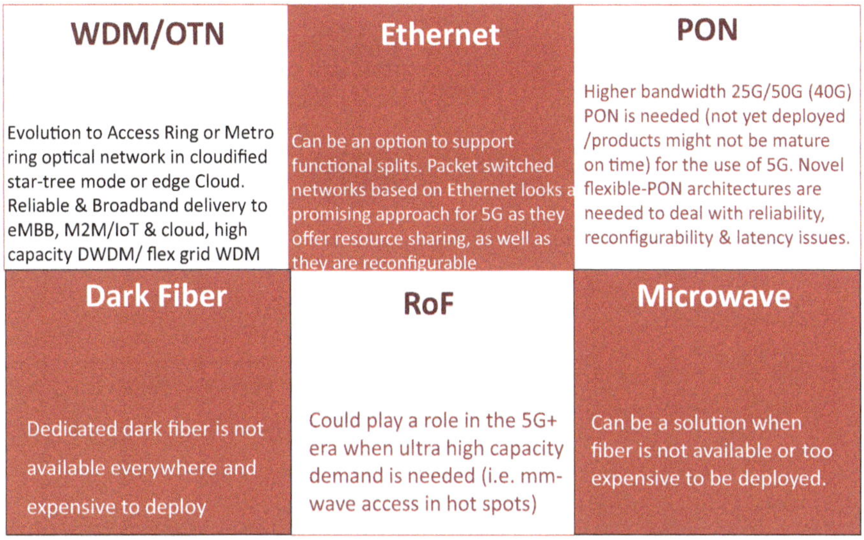

Figure 5.7. 5G optical fronthaul, highlight characteristics of core and optional solutions.

risk technologies such as radio over fiber (RoF) can be considered in the long term (5G+) as digital RoF (CPRI-like) cannot cope with millimeter-wave technology (ADC, etc) and RoF needs only the BW required by the wireless channel. In all cases, the introduction of WDM technology will need resource sharing. Putting as many technologies with similar requirements together will increase the production volume and thus reduce the cost.

5.2.3.2 Photonic enabled technologies including principal devices and photonic processors

In order to cope with the 5G revolution demands from an optical communications viewpoint, breakthrough technologies in optical interfaces and components will be needed. In particular, advanced technologies such as silicon photonics can be the cost efficiency enabler for C-RAN based on optical fronthaul and WDM. Silicon photonics (SiPh) offers compatibility with the CMOS technology and will enable the low cost use of optical transport in 5G access. SiPh devices such as C-band reconfigurable optical add–drop mux (ROADMs), or 25 G externally modulated lasers (EMLs) able to support the 25 G rates, and in the future 50 G and 100 G must be available. ROADMs based on SiPh will lead to cost reduction compared to the conventional ones. DCI, OTN, 5G and PON all share 25 G requirements (figure 5.8) so SiPh devices should target the speeds of 25 G and beyond and will enable a low cost solution for C-RAN based on the optical fronthaul. Moreover, graphene can transform SiPh to a low cost, CMOS compatible, outperforming III–V equivalent platform, but great investment is needed to transform prototypes to industrial products.

The frequency bands for 5G, in particular 5G+ apart from the 2–6 GHz, are located in the microwave and millimeter-wave (mmW) centralized at 18.6 GHz and 28.6 GHz or 56.8 GHz and 90 GHz. The free-license band of 7.0 GHz of the

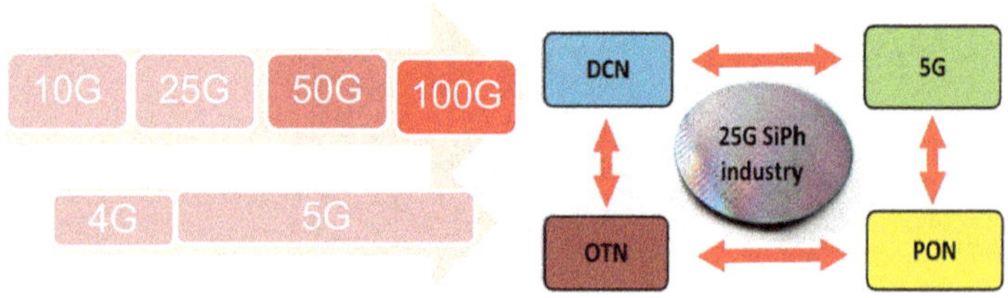

Figure 5.8. Component requirements and resource sharing drive optical technology.

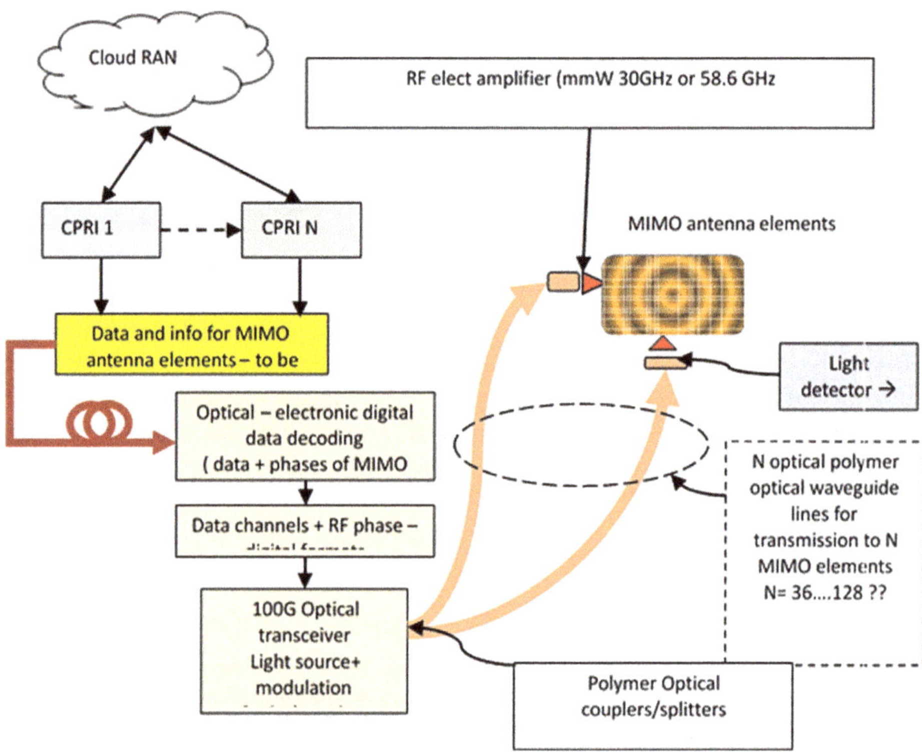

Figure 5.9. Integrated photonic processor in an OTTA optical to the mmW MIMO antenna.

58.6 GHz data channels can be delivered to mmW antennae, which are multiple input multiple output (MIMO) integrated horns, via an optical guided medium and converted to the mmW electronic frequency region and amplified to drive mmW MIMO antenna elements.

Connecting devices range from massive capacity premises such as education campus data centers and direct access by academic departments, machine-to-machine communications, fiber to the premises (FTTP), and fiber or optical to the antenna (OTTA), as shown in figure 5.9, in which the data communication link delivers data to the mmW antenna elements where the optical signals are received and mixed with a local oscillator laser in the photodetector. The frequency of the LO is tuned to a frequency difference with respect to that of the data carrier of mmW center frequency, e.g. 58.6 GHz. This results in opto-electronic conversions of optical signals to the mmW band. The mmW power amplifier delivers the signal to the antenna element, a millimeter size horn antenna embedded in a printed circuit board. This is the optical to MIMO antenna device for a mmW band wireless carrier. This scalable design can have a bank of receivers and mixers and mmW RF electronic drivers. A bank of photonic phase shifters is used to generate a pattern of phase delay so that beam steering can be implemented. These phase shifters must be modulators in an integrated optic structure so that fast and high speed beam steering can be performed. Furthermore, these OTTA are passive optical networks for distribution and hence enable sharing of capacity and cost. The integrated photonic processor can be implemented in silicon photonics to keep the costs sufficiently low. Beam steering phase modulators can be embedded using p–n junctions by ion implantation, operating in junction reverse mode, whose speed can be in tens of GHz.

With the 7 GHz bandwidth of the free-license band and 500 MHz per sub-band for data allocation, high level complex modulation can be used as in common RF wireless so that more than 10 Gbps can be delivered to MIMO antenna elements. With an antenna gain of about 22–0 dB for a 56 × 128 element array of the MIMO antenna we can deliver more than 1.0 Tbps capacity covering an area of radius of 500 m by beam steering to clients of a crowded stadium.

5.2.3.3 Beyond 100 G low cost PAM4 systems for 5G

Low cost modules for DCN are critical for lowering the cost of DCNs, particularly in 5G distributed DCN. One way to lower the cost of a high speed module is to use components for 10 Gbps but for 100 G (25 G × 4 lanes) but equalized by an MLSE algorithm implemented in the DSP. This sub-section presents the technique of this low cost approach.

The system set-up is shown in figure 5.10. The 10 G components can be from manufacturer 1 (Sumitomo), manufacturer 2 EA modulator integrated with DFB laser (EML), manufacturer 2 direct modulation laser (DML) which is a laser modulated by direct driving laser current, and another DML version from manufacturer 2. These components are frequently used in the 10 G transmission link and operating in the O-band (1310 nm spectral window). At the receiver side an optical receiver with a bandwidth 40 GHz 3 dB passband is used. This receiving sub-system is composed of a high speed PD in tandem with a TIA of a spectral noise current density of 40 pA/sqrt(Hz) and a trans-impedance transfer power gain at a mid-band of 150 V W^{-1}.

As observed, the effective bandwidth can be 27 GHz with 10 dB and a higher SNR. The number of bits per symbol that can be used varies from two to six over the 27 GHz band. Thus the order of higher modulation formats can be employed accordingly in the OFDM or DMT. Small impedance mismatch causes some degradation of the SNR, as observed in figure 5.11.

The digitally modulated sequence is generated and output to drive the transmitter optical assembly (TOSA) via a DSP platform whose sampling rate is 100 GSa s^{-1}. 112 Gbps is successfully transmitted over a 2 km SMF with a sensitivity of −2.5 dBm

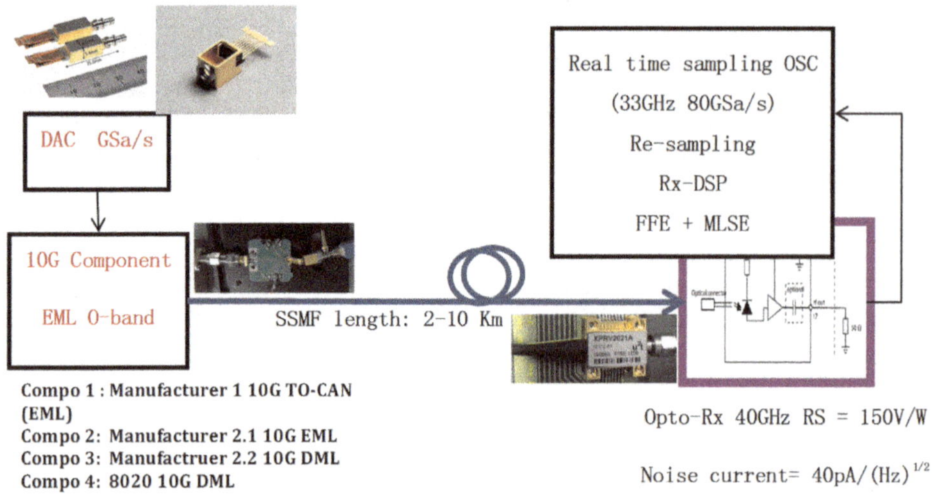

Figure 5.10. Low cost 100 G transmission system by 10 G components.

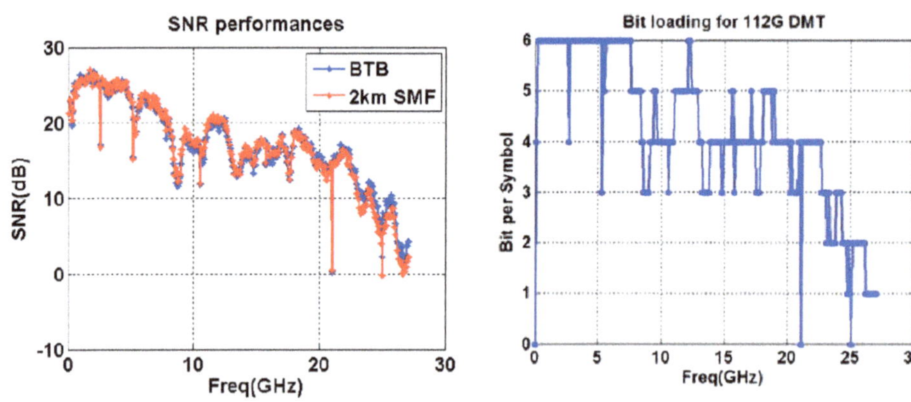

- The highest used frequency is around 27GHz
- Power fading possibly caused by mismatch between the PCB and component?

Figure 5.11. Electrical signal-to-noise ratio (SNR) and the bit per symbol versus frequency to 30 GHz.

which is only a 0.2 dB power penalty after 2 km SM. Improvement is required to reduce the error-floor and the highest bit rate is 120 Gbps for both back-to-back and over a 2 km length of SSMF. Table 5.1 shows the ROP and capacity performances for the Sumitomo TO-CAN 10 G component transmitted over 2 km of SSMF. Physical components are shown in the insets in figure 5.12. For the EML and TOSA, the results can be seen in figures 5.13 and 5.14, respectively. Figure 5.15 shows the DMT BER performance for a DML based transmission system. Finally, table 5.1 gives the comparative performance metrics for the three manufacturer's components.

Table 5.1. Summary of performances of different TOSA EML and DML components of different manufacturers (Sumitomo and manufacturer 2(a) and (b)).

	Sumitomo EML	Manufacturer 2(a): EML	Manufacturer 2(a): DML	Manufacturer 2(b): DML
B2B ROP (dBm)	−2.7 @ 112 G	−7.6 @ 112 G	−1.8 @ 84 G	−2.8 @ 84 G
2 km ROP (dBm)	−2.5 @ 112 G	−7 @ 112 G	0.2 @ (84 G)	−3 @ (84 G)
5 km ROP (dBm)		−6.5 @ (112 G)		
10 km ROP (dBm)		−9.5 @ (84 G)		
B2B capacity (Gbps)	120	162	108	101
2 km capacity (GBps)	120	147	92	97
5 km capacity	—	130	—	—
10 km capacity	—	99	—	—

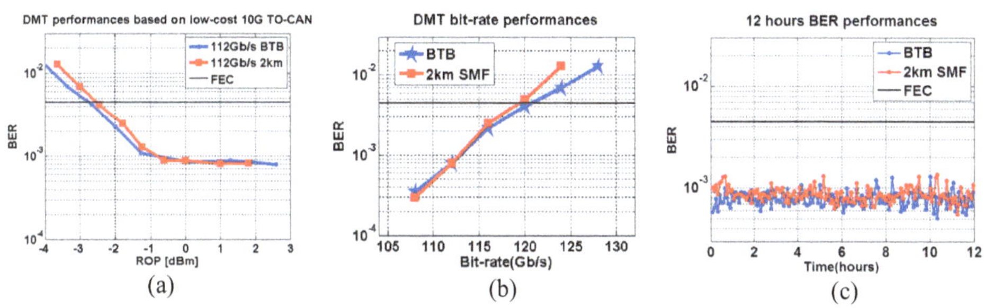

Notes: 112Gbps is successfully transmitted over 2km SMF with a sensitivity o f -2.5dBm; Only 0.2dB power penalty after 2km SMF; Improvement is required to reduce the error-floor; and The highest bit rate is 120Gbps for both B2B and 2km cases.

Figure 5.12. Sumitomo TO-CAN TOSA: BER versus (a) receiver sensitivity (dBm) and (b) bit rate from 105 to 130 GBps. (c) Stability of BER with respect to time in hours. B2B = back-to-back.

5.2.3.4 PAM4 150 Gbps by band-limited components

It is understood that the bandwidth of electronic and integrated optic devices, in particular Si photonics, is limited to around 30–40 GHz. So the total effective bit rate can be limited to around 80 Gbps for an NRZ modulation format. This section, however, attempts to describe a method for NRZ transmission but with PAM4 duobinary (PAM4 DuoB) using electrical filtering with no optical phase in association with DSP to demonstrate bit rates more than 150 Gbps. The experimental transmission system is shown in figure 5.16.

(a) A laser is coupled to a dual electrode Mach–Zehnder interferometric modulator (MZIM) whose 3 dB bandwidth is about 32 GHz. The laser linewidth is about 2 MHz. The MZIM is driven by two RF amplifiers fed by the analog signals output of the DAC of an arbitrary waveform generator (AWG). More details are given in the inset of figure 5.16(a). The transmission link is a 2 km SSMF. The transmitted signals are received by an optical receiver whose bandwidth is limited by the 3 dB bandwidth of

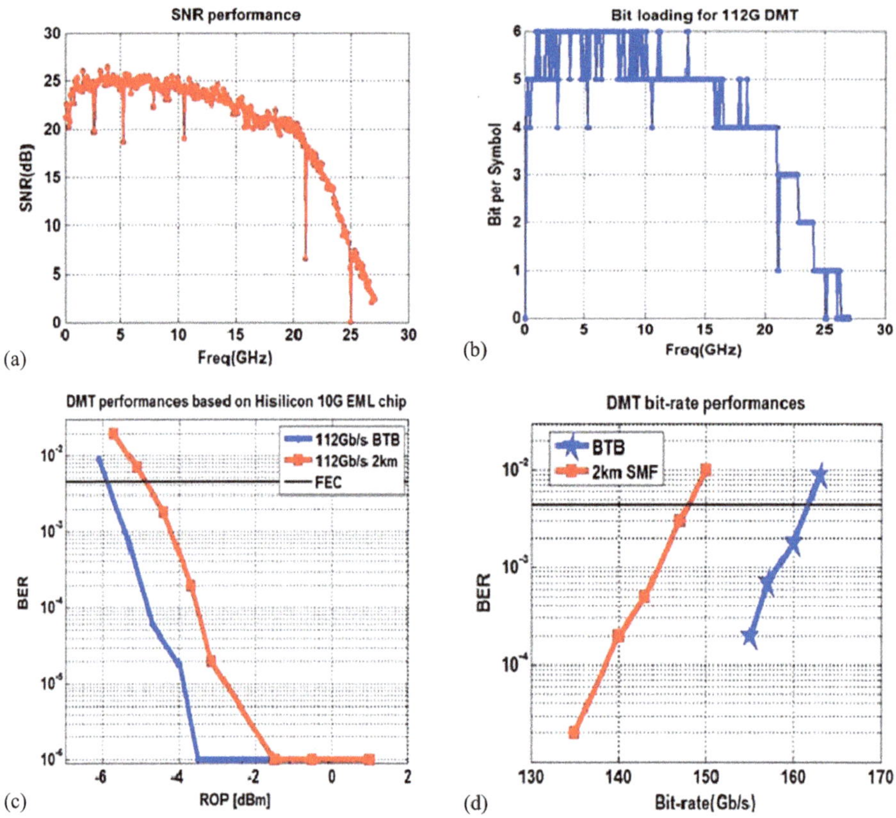

Figure 5.13. Manufacturer 2 EML: (a) SNR and (b) bit-loading of 10 G EML chip. The highest used frequency is around 27 GHz; narrow power fading is caused by clock leakage. Manufacturer 2 EML: (c) ROP and (d) capacity performances for 2 km SMF.

34 GHz of the trans-impedance amplifier (TIA). The signals are sampled and processed in the DSP at a sampling rate of 90 GSa s^{-1}.

The PAM4 duobinary eye diagram shown in figure 5.16 indicates a seven level eye, confirming the half band filtering of the PAM4 eye. These eyes do not seem to be affected after the transmission over 2 km SSMF. The signal sequence is sampled, processed and recovered by maximum likelihood equalization (MLSE) and then measuring BER with respect to the variation of the receiver sensitivity by varying the attenuation coefficient of the variable optical attenuator (VOA). An FEC is used and the BER (1e-3 to 1e-4) is set at the level so that 7% FEC and 20% soft FEC can be allowed, as indicated in figure 5.16.

(b) The transmission distance is varied from back-to-back to 2 km with a 1 km step. The sequence bit rate is 180 Gbps, that is, a baud rate of 90 GBd by the Nyquist sampling theorem.

In summary, the duobinary filtering in the electrical domain reduces the total effective bandwidth of the PAM4 and thus permits minimum

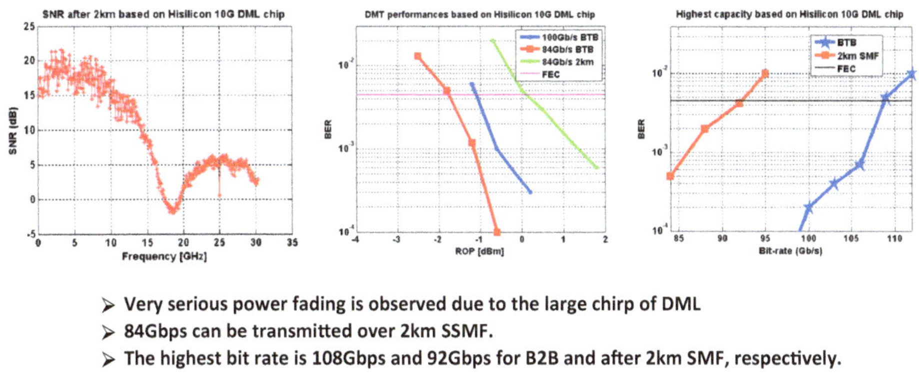

Figure 5.14. Manufacturer 2 10 GHz DML TOSA: Performances for 2 km transmission (a) frequency response of SNR, (b) BER versus receiver sensitivity and (c) BER versus bit rate—modulation DMT.

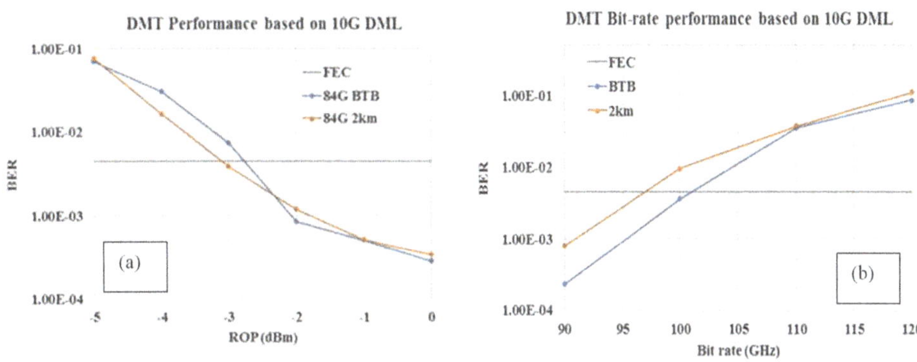

Figure 5.15. Manufacturer 2 DML: BER versus (b) receiver sensitivity (power input) and (b) bit rate. Capacity performance for 2 km using manufacturer 2 DML.

distortion of the signals and the DSP can recover the original PAM signals with minimum reduction of the sensitivity of the receiver. Indeed, this bit rate can be increased over 200 Gbps if the bandwidth of the system components can be enhanced to the upper limit of about 40–45 GHz. Alternatively, the OFDM or discrete multi-tone (DMT) technique can be employed to increase the transmission rate. The disadvantage of OFDM is the power consumption as it is a bit high due to its high sampling rate. It should be noted that the power dissipation is proportional to the cubic function of the sampling rate.

5.2.3.5 Security aspects and transmission technology for secret keys in the co-transmission of massive data transport

5G radio needs to be flexible, dynamic and have the ability to handle high capacity. A converged software defined networking/network function virtualization (SDN/NFV) based optical transport solution, will support a large number of services in

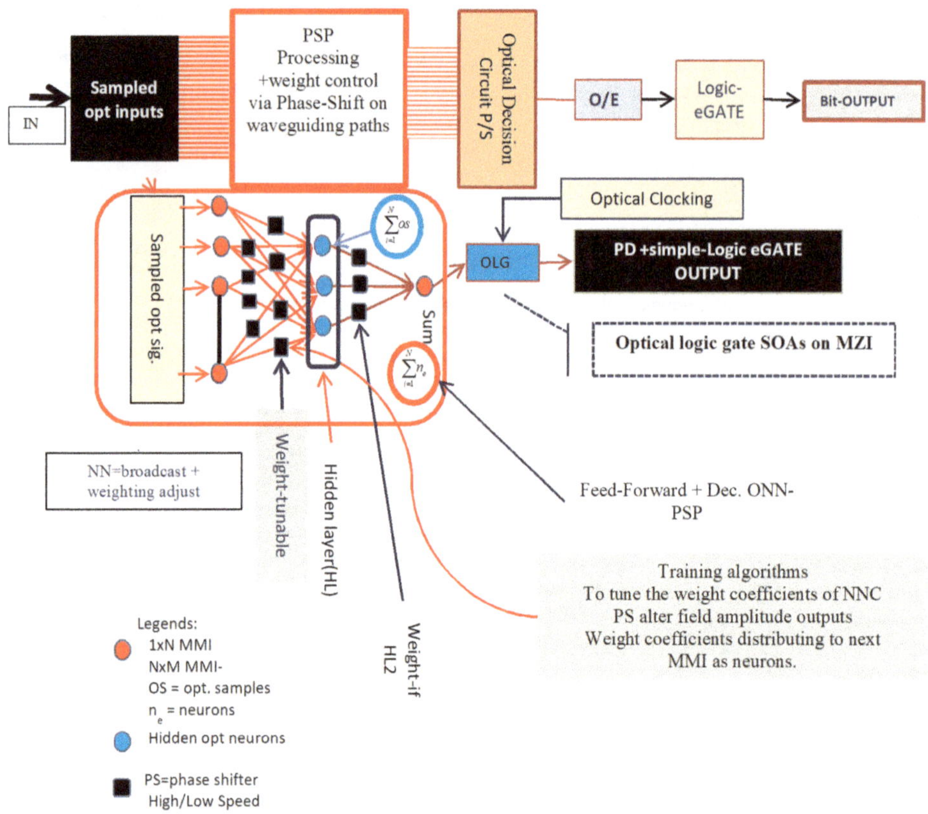

Figure 5.16. Demonstrated 180 Gb s^{-1} DB-PAM4—the bit rate is limited by the AWG (maximum 90 GS s^{-1}), MLSE to recover the pulse sequence.

parallel on a dynamic network operation to significantly reduce costs. SDN will allow the separation of control and data planes to reflect the integration of core, metro, access and 5G on a single SDN-based network platform. However, it is likely that full-scale deployment of an SDN-base architecture will take more time than expected, limited by capacity, compatibility, security and transformation cost issues, as well as orchestration and management which needs to be optimized. In particular, secure communication will receive more attention in SDONs, however, classical key distribution is subject to attack algorithms, computational power consumption and in the long term will be more relevant with the rise of quantum computers. Quantum key distribution (QKD) can be considered as an additional layer of security and can be used in the 5G backhaul or the fronthaul optical network. Continuous variable QKD (CV-QKD) can integrate quantum key distribution using a legacy network (such as OTN) without the need for dark fiber infrastructure, with the additional advantage of using standard devices with minor modifications. Optical-SDN in 5G backhaul/fronthaul, i.e. decoupling of control and data layer, provides a unique case for QKD. Quantum keys can be used to secure control layer, and integration of QKD into a SDN network can greatly reduce the CapEx as the transmitter can be shared among different receivers, thus fewer QKD devices will be needed. A relevant SDN-enabled QKD prototype has recently been demonstrated by Telefonica in Madrid.

5.2.3.6 Summary

In this section we discussed and demonstrated a 5G transport scheme that can be implemented as an evolutionary step to support future DCs. The successful realization of this transmission link demonstrates the potential of the photonic integrated circuits to be implemented in 5G infrastructures.

5.3 Photonic signal processors

Integrated optic technologies have progressed tremendously over the years, in particular since the beginning of the millennium and after the 'dot-com' problem. The baud rate has now reach 100 GBd and even higher (200 Gbd) via time multiplexing. Thus under coherent transmission and reception, techniques can offer nearly and higher than 1.00 Gbit s^{-1} for metro and short distance or inter connections in data centers.

At these baud rates the operation of the transmitter and receiver can be better by operating everything in the photonic domain, including the cross connect integrated circuits as the electronic switching speed has reached the limit and the power consumption will increase drastically when such baud rates and bit rates are processed in the electronic domain.

This section gives an overview of photonic signal processors (PSP). By photonic processing we mean the manipulations of photons in both the active and passive domain in coherence while all-optical waveguide paths indicate highly passive waveguides for conections and interconnections of photonic devices. A schematic diagram of the photonic processor is given in figure 5.16.

5.3.1 Generic deep neural network learning photonic signal processor (DNNLPSP)

In practice the all-optical signal processing is workable provided that one can determine exactly the optical guided path. The OFT processor is a linear processing system with minimum use of optical decision circuitry. This section briefly presents the photonic signal processor using optical neural networks with deep learning, in which an optical decision maker circuitry for decision feedback and tuning the optical neural network are used so that convergence can be achieved. This section provides a proposed photonic signal processor whose structure consists of a central optical neural network (ONN) in association with optical sampler and optical decision circuitry so that feedback can be used to tune the weighting coefficients of the ONN. The summation and subtraction functions that can be implemented are given in the optical domain by the multi-mode interferometer (MMI) and an integrated phase shifter so as to change the coupling coefficients, hence the variation of the total waves at the output port of the MMI. Table 5.2 shows the abbreviated terms used in this section.

Table 5.2. Abbreviated terms used in this section.

FF = feed forward	SiIP = silicon integrated photonics
FB = feedback	SOI = silicon-on-insulator
DFB = decision feedback	SOA = semiconductor optical amplifier
ODFB = optical decision feedback	OLG = optical logic gate
O/E = optical to electronic conversion	MZI = Mach–Zehnder interferometer
eDSP = digital signal processing (electronic)	NOT gate = optical inverter
MCU = micro-control unit	AND/NAND = logic gate AND and not AND
ONN = optical neural network/networking	XOR = exclusive OR gate
OSP = optical signal processing	OC = optical coupler
PSP = photonic signal processing/processors	
S/P = serial-to-parallel conversion/converter	
P/S = parallel-to-serial conversion/converter	

The motivations for processing in the photonic domain can be stated as follows. (i) Electronic DSP at an ultra-high sampling rate, e.g. 200 GSa s^{-1} for 100 GBd, would require very high power consumption and the limitations in operating speed, in terms of which electronic DSPs are facing considerable difficulties, in particular at 200 GSa s^{-1}, and ultra-high power consumption (note: power consumption in electronic sampling increases by a cubic factor of the sampling increase factor). (ii) The processing speed in the optical domain can be much faster than that by the electronic DSP on the condition that an accurate delay in the optical domain can be fabricated for a sampling speed of 1–2 ps delay, which can currently be implemented in Si photonics technology. (iii) The advances of integrated photonic devices, in particular in Si photonics based on silicon material, waveguides, and active as well as passive device structures for modulation, routing, splitting, etc, make it possible to implement several optical functions.

This section proposes a generic PSP structure using an ONN and FF decision optical logic circuit. The idea is to extend the basic structure to include optical feedback from the output of the OLG. Such optical feedback will minimize the complexity of the training scheme and accelerate the convergence of the ONN to the desired output so that the determination of the signal level can be obtained. We refer here to figures 5.17 and 5.18 and provide a more detailed description of the operational principles in the next sub-section.

The optical data which are derived from either transmitter (Tx) or receiver (Rx in the optical domain) or even the outputs of all-optical interconnectors or switches. The optical data sequences are sampled and converted to parallel form via the use of optical delay lines of different delay times in optical guided wavelines. This block is indicated as 'Optical S/P'. Once the data channels are in parallel form these optical paths are fed into a photonic signal processing unit which is a sub-system of the PSP. The photonic processing is implemented using an optical neural computing network.

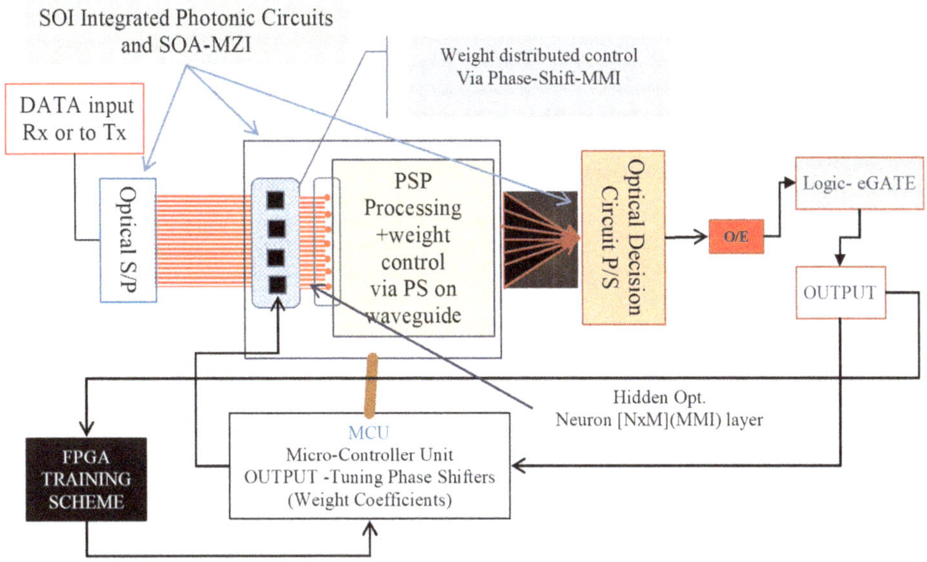

Figure 5.17. Schematic of PSP principles. PS = phase shifter; MCU = micro-controller unit; eGate = logic gate in the opto-electronic domain; S/P = serial-to-parallel conversion; Rx/Tx = receiver or transmitter; SOI = silicon-on-oxide; MMI = multi-mode interferometer.

All manipulations are implemented on SOI, which are the silicon photonic devices fabricated in a Si thin film guided medium which can be activated via the electro-optic effects in the C-band silica fiber. These SOI integrated optic devices are now quite popular in the field of integrated photonics. The operation of this optical neural network (ONN) is described later in this section. In principle, optical neural computing is composed of two main functions, one is to distribute one path to all other paths where the weightings or magnitudes of optical fields are differently controlled. These weightings are controlled and adjusted continuously until the convergence of a target level which is decided by an optical decision circuit, which derives from the summation of all the optical paths to a nonlinear optical opto-electronic sub-system, the O/E converter and the e-gate or electronic gate which is set at a level for decisions of 'high' or 'low'.

The ONN can be trained via a field programmable gate array (FPGA) sub-system which output electronic signals to a micro-control unit (MCU) whose output signals are in voltage or current levels feeding to the electrodes of the phase shifters (PS) to tune the optical phase via the ferro-electric effects, or electro-optic effects if higher speed is required. These phase shifting actions in turn tune the coupling coefficients of the outputs of the MMI. This is the physical insight of the MMI and the central operating principles of the ONN. The control can be very fast with specific processors whose sampling rate can be in the pico-seconds and whose electro-optic phase modulators use integrated Si photonics.

We now summarize the operation of the PSP and ONN in the following sections.

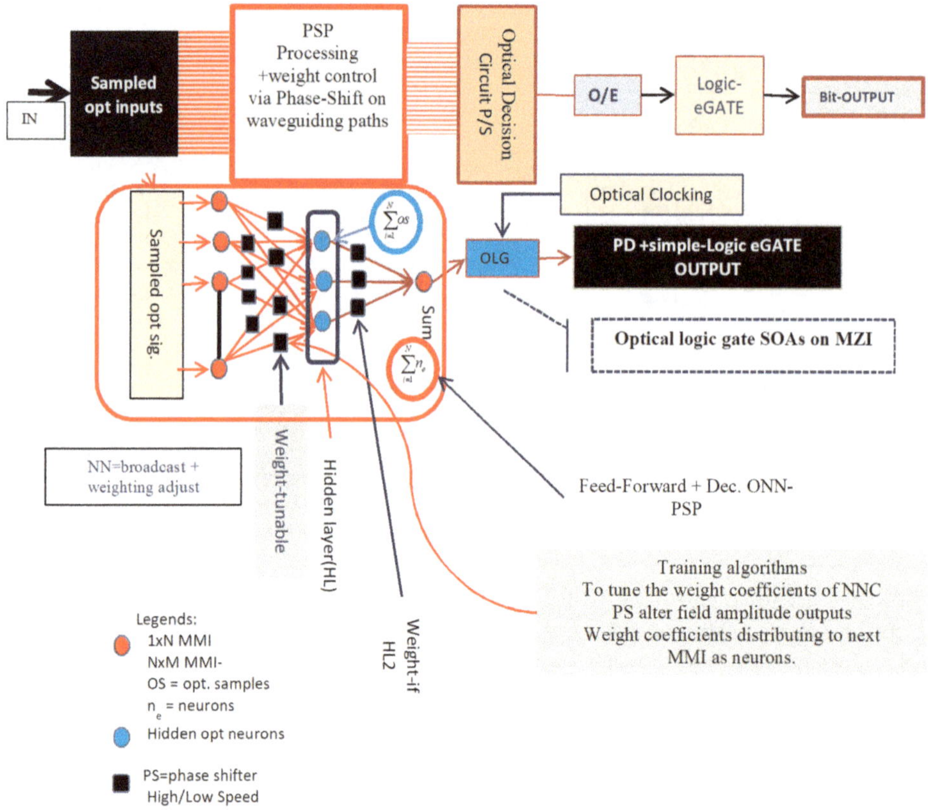

Figure 5.18. Schematic of a PSP system by optical neural network computing (ONNC).

5.3.2 PSP operating principles

Schematic of generic PSP system principles:

- Equivalent DAC in optical domain: Referring to figure 5.19, the optical sampling concept and optical serial-to-parallel (S/P) conversion without memory and a first-in first-out (FIFO) dynamic action occur for sampling the optical signals, thence generating the optical discrete samples.
- PSP of parallel sampled optical inputs: optical neural network computing algorithms
- Neural net computing (NNC) optical algorithms: feed forward NN (FF-NN) w/o decision, FF-NN-decision, FF-NN-decision feedback (FB), optical reservoir NN computing system/chip.
- Optical decision circuits: SOA-MZI.
- Algorithms of deep NN learning (that is, with back propagation) and possible applications to PSP: to be applied for patents after the approval of the Patent Committee of this idea. Such as those listed in the third point, namely FF decision FB equalization PSP, adaptive FF equalization PSP, etc.

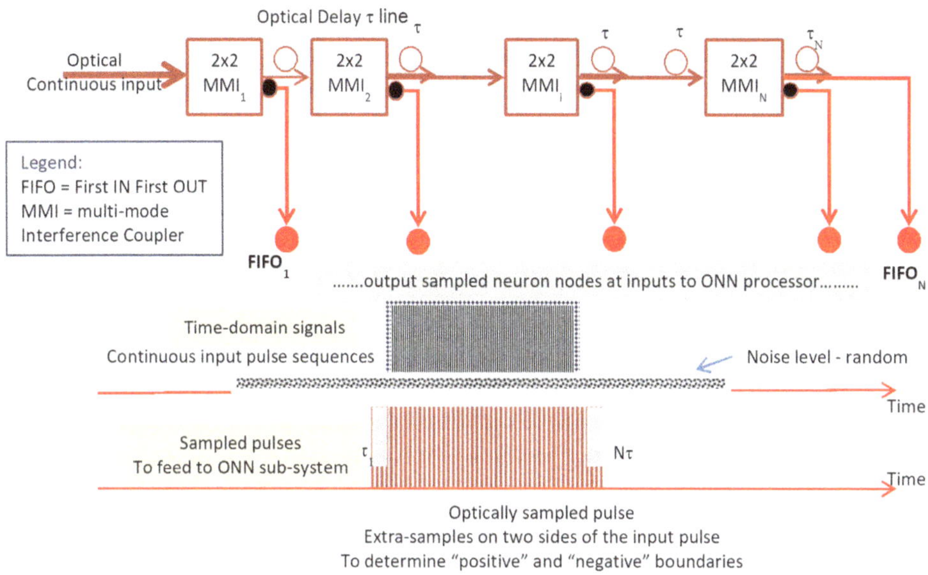

Figure 5.19. Optical analog sampling by MMI and delay lines, equivalent to a sampler in DAC/ADC. FIFO = first-in first-out.

5.3.3 ONN operating principles

- Input optical signals (including embedded noise) are converted to parallel by optical sampling (the optical sampler—OS and serial-to-parallel conversion) in the optical domain by using optical delay and split and coupling out by an MMI coupler—similar to transversal filtering in the electrical domain. So these samples are equivalent to digital optical signals but in optical analog features. The principles of operations are given in the following dot points.
- The optical sampled signal waves are fed into a PSP processor consisting of hidden optical neuron layer(s) or an MMI with controlled phase via the MCU so that the weighting coefficients can be varied/tuned. (See the generic structure of optical neural networking for PSP, given in figure 5.18).
- After optical NN computing by various processing algorithms the outputs of optical neurons are fed into a neural summation and then fed into an optical logic gate (OLG) so as to determine that the processed inputs can be a '1' or '0' for NRZ or '00', '01', '10' or '11' of PAM4 signals for one directional signal.
- OLG can consist of two semiconductor optical amplifiers (SOAs) embedded in an MZI with three coupled inputs which are allocated for optical signals, control signals and optical clocking signals.
- In the case of complexed signals then there would be another $\pi/2$ optical phase shift on the split optical path ($\pi/4$ PS) to another ONN and optical logical gate.
- Training schemes of different ONN structures are to be conducted so that fast convergence of the optical processing can be achieved. A look-up table could

be used if compensation of nonlinear impairment effects is required for fast convergence to the targeted output.

- Our proposed PSP system is highly stable as no self-optical feedback, which is a positive feedback, or oscillation can happen. But without self-feedback our PSP system is memoryless and highly stable.

The operation of optical logic gates can be described as follows (figure 5.20):

- OLG consists of two SOAs embedded in the optical paths of an MZI. Thus the operation relies on the phase changes in the SOA under signal and control signal inputs as well as the third input which is considered as a clocking signal as shown in figure 5.19.
- Data signal 1 and data signal 2 are coupled into the MZI paths via the optical coupling region. Two control signals can be used to feed into the MZI. First the control signal can act as an ENABLE to allow the output of the MZI-SOA to operate, or a clocking optical signal can be used to clock the data sequence out.
- The input signal comes from the total sum of all the optical samples which are processed by the ONN. If the signal amplitude is large enough to compare logically with the reference optical signal (signal 2) then an XOR or AND operation can be performed. This is the nonlinear optical logic operation as required for decision making in the optical domain.
- The novelty of the use of this OLG is that it simplifies the optical decision circuitry or uses only one optical device, unlike the other two recently reported schemes [2, 3] in which a PD array must be used.

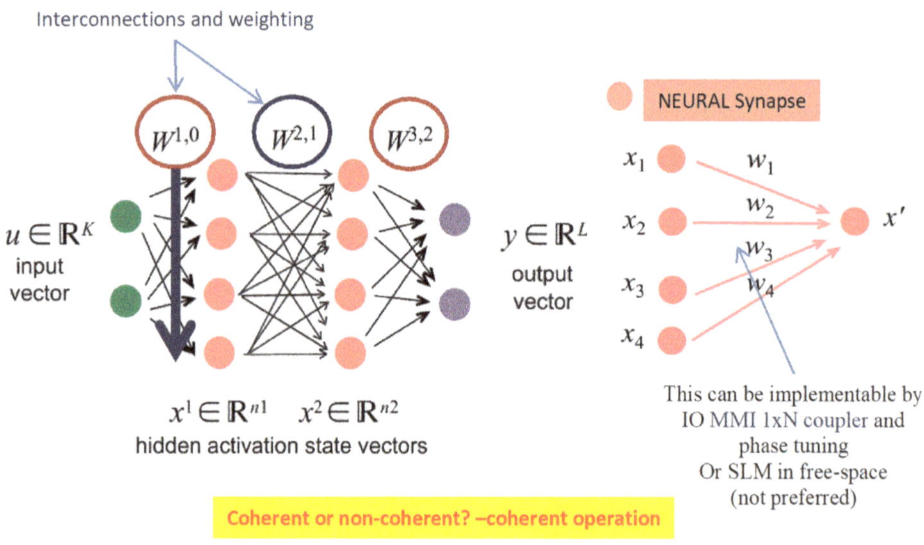

Figure 5.20. Generic model optical neural networks—functions and technology. SLM = spatial light modulation.

- Many logic gates can be formed by this structure. The reference optical signal 2 can be set after the training scheme as indicated in previous slide. These OLG will be used in the coming patent idea submissions.
- The speed of logic operations can now reach 20–30 GHz and is expected to be higher depending on the speed of the phase changes of SOA.
- Possible logic operations using MZI-SOA: XOR, AND, NAND and NOT.

References

[1] Hillerkuss D *et al* 2011 26 Tbit/s line-rate super-channel transmission utilizing all-optical fast Fourier transform processing *Nat. Photonics* **5** 364

[2] Silicon chip distributes optical signals for potential use in neural networks *Photonics.com*, Aug 2018 https://photonics.com/Issues/Photonicscom_August_2018/i1040

[3] Chiles J, Buckley S M, Nam S W, Mirin R P and Shainline J M 2018 Design, fabrication, and metrology of 10 100 multi-planar integrated photonic routing manifolds for neural networks *APL Photonics* **3** 106101

Lightning Source UK Ltd.
Milton Keynes UK
UKHW050745040721
386574UK00003B/143